No. 8 Re-wired

202 New Zealand Inventions That Changed the World

JON BRIDGES & DAVID DOWNS

PENGUIN BOOKS

Contents

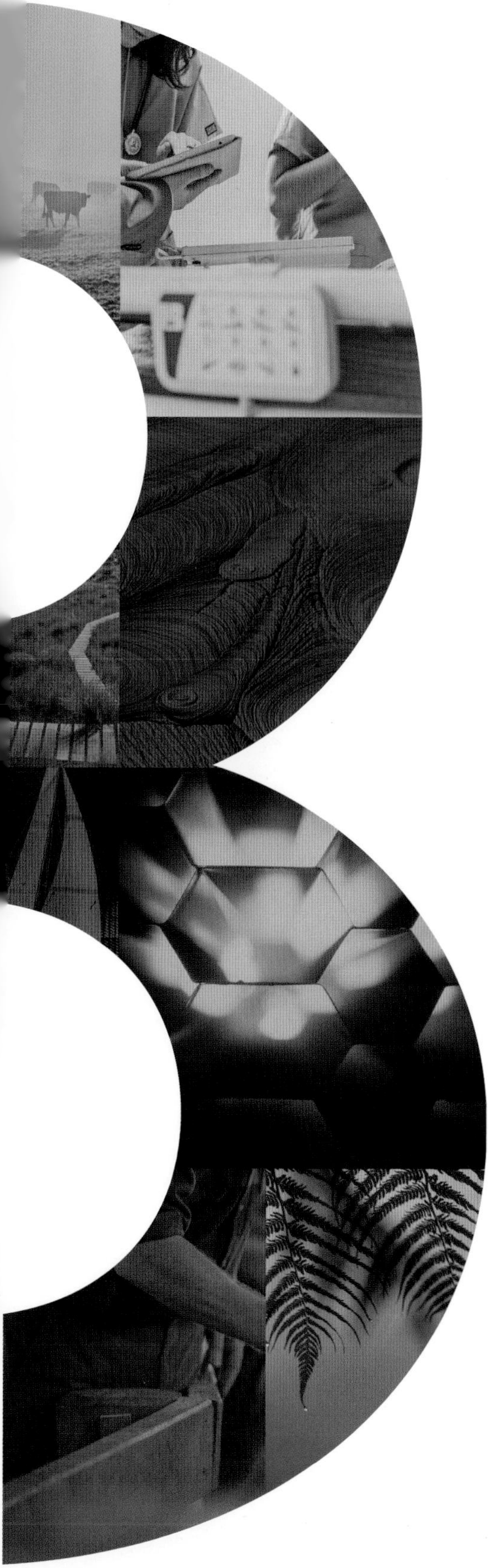

‘We haven’t the money, so we’ve got to think.’

LORD SIR ERNEST RUTHERFORD
Nobel Laureate, farm boy from Nelson

Foreword

SIR PETER GLUCKMAN
Chief Science Advisor to the Prime Minister of New Zealand

This book is a compendium of creativity. It tells the intriguing, impressive, often surprising and sometimes frustrating stories of innovative New Zealanders. But it also carries a deeper message – it extends the concept of innovation well beyond the technological to also include stories about social and methodological innovation as well.

It talks about great New Zealanders who made their contributions not just in New Zealand but after they had moved beyond our borders. The names resonate, and include both the research giants and the quietly curious on whose shoulders others have stood: Pickering, Rutherford, Phillips, Wilkins, and more – just imagine had some of them been able to do their work in New Zealand!

Entertaining as this book is, authors Jon Bridges and David Downs highlight a particularly frustrating problem in their introduction: they challenge us to stretch beyond the received wisdom and clichéd mythology in the way we think about ourselves. It is this challenge that makes this book so compelling – we might view the title of this book as a national goal. Can we re-wire ourselves and use our innovative individualism to national advantage?

As New Zealanders, we see ourselves as inventive, innovative and independent, and these 'national character' traits are reinforced and perpetuated in the story we like to tell (ourselves and others) about being able to fix or build anything with No. 8 Wire. However, the question posed throughout this book is whether tying our identity to this national myth is holding us back. Are we simply denying reality?

It is too easy and simplistic to assume we are limited by our geography, and while it does create some costs and logistical challenges, modern information and communications technology have changed the equation. High-quality food and eco-tourism remain the mainstays of our economy and, relative to other outputs, these are easy to sell and are in demand.

Our geographical remoteness and social history have certainly imparted a strong sense of individualism and practical know-how. While these values might have worked well in past decades, in a globalised economy, they will simply no longer sustain us into the future. Rutherford's famous comment (opposite) has become an excuse rather than a solution. Simply being clever Kiwis, without the investment required to turn our ideas into a new or better product or method, cannot succeed.

It is only when we develop a more complete innovation ecosystem and address some of the issues Bridges and Downs raise that New Zealand will really emerge as a South Pacific tiger economy. We cannot pretend we are already there. But things are changing: politicians are now talking in terms of R&D as an investment not a cost; there is a more mature understanding of the policy levers that promote investment and thus innovation; and some exciting examples of science-based innovative companies reaching to the world are emerging. The Tall Poppy syndrome, which is probably another remnant of our cultural past, is beginning to fade.

No. 8 Re-wired sheds light on the many past opportunities New Zealand has had. And if past performance is a predictor of future performance, we still have a lot to look forward to. But this will only happen if we can divest ourselves of the national myths that hold us back, while retaining and indeed wielding those that inspire, unite and challenge us.

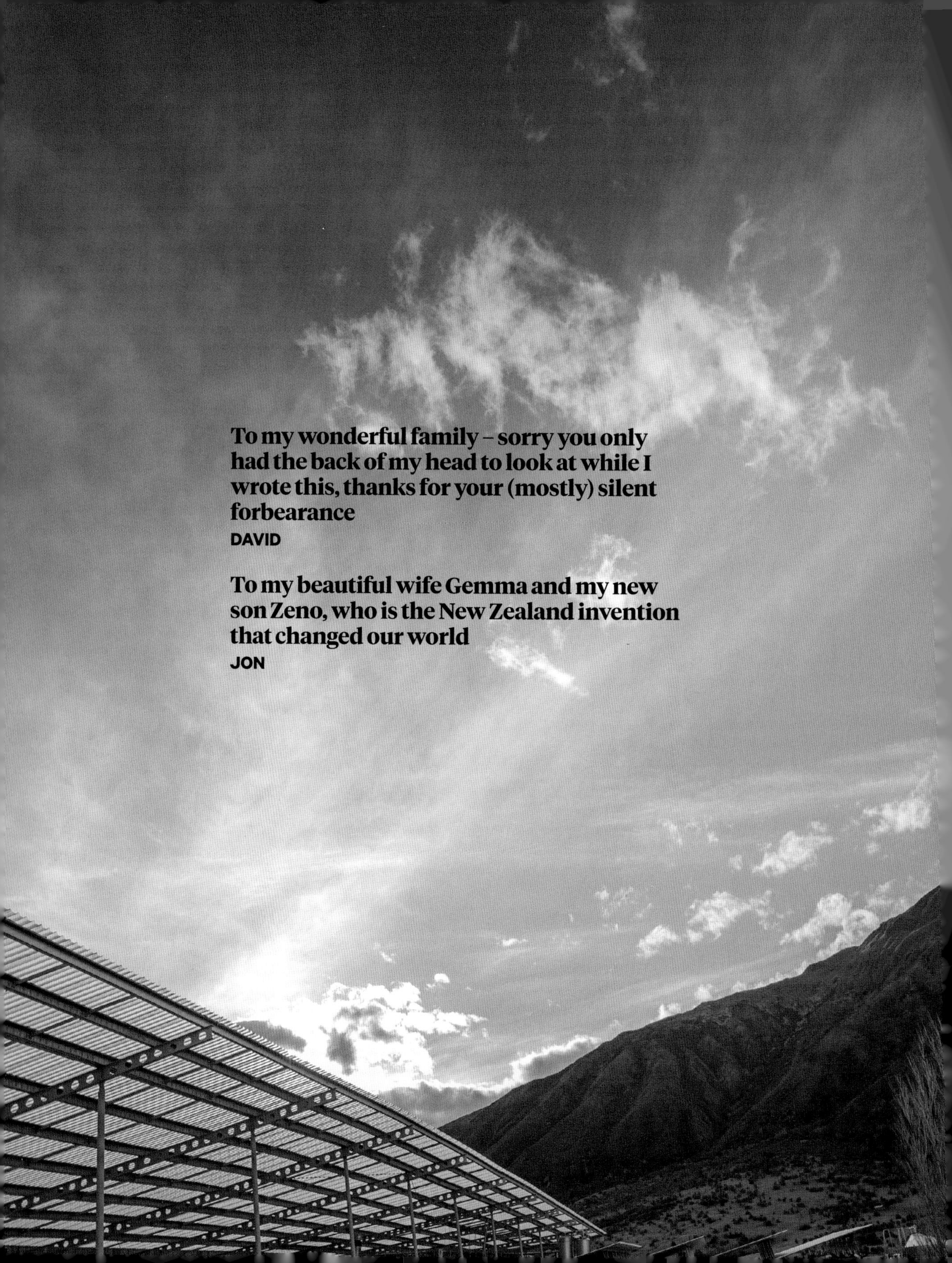

To my wonderful family – sorry you only had the back of my head to look at while I wrote this, thanks for your (mostly) silent forbearance

DAVID

To my beautiful wife Gemma and my new son Zeno, who is the New Zealand invention that changed our world

JON

Introduction

We Kiwis think of ourselves as an inventive and innovative nation. We even have a phrase for it. Our 'No. 8 Wire' mythology celebrates the inventive Kiwi fashioning a novel solution out of necessity and with few resources other than cunning, pragmatism and determination (and possibly fencing wire).

As the entries in this book will attest, our past is populated by amazing people with incredible, often world-changing ideas. This book is a celebration of our great ideas and the impact that we've had on the world.

But the 'No. 8 Wire' image is also *not* a true reflection of who we are – and in fact is a dangerous paradigm for the future. New Zealand is no longer a world leader in solving intractable problems with clever ideas. Our levels of innovation now trail our peers', and we the authors contend that some of the core attributes of our culture, identity and psyche need to mature for us to ensure a lasting relevance and impact on the world. So this book also aims to discover what innovation looks like in the modern world and how we can improve our performance by moving from No. 8 Wire to No. 8 Re-wired.

About this book

We wrote this book's precursor *No. 8 Wire: The Best of Kiwi Ingenuity* in late 1998 in what was, even then, a different world. A few years of living overseas – with the perspective that brings – coupled with increased opportunities to see New Zealand innovation up close and in a modern context led us to believe that the topic needed reconsidering. This update has given us the chance to update the book with 15 years of new inventions, and to highlight the challenges of working the same way in a changed world.

In this edition, expect to see an extreme variety of inventions: the classics that everyone knows of; the folklore and legends we hold onto; new things from long ago that we are only now claiming ownership of; things that made millionaires and things that ruined people completely; and a number of exciting new ideas – some hard for the layperson to understand – that may well change the world.

In choosing what to put into this book we have sometimes taken a broader view of an invention than the dictionary definition might allow. An invention is typically thought of as a gadget, a widget, a 'thing', a machine or at least a tangible object that you can hold, ride, see, catch or put in your pocket. But there are some things that are made brand new and spread around the world that aren't necessarily like that. Ideas like democracy *for all*, processes like the splitting of the atom, and discoveries like the structure of DNA join other categories in this book such as foods, fads, theories and equations that were all startlingly new and changed the world.

Defining what is really an invention versus an evolved version of someone else's idea was at times difficult. As a guideline, we developed some basic criteria for deciding if something is a great New Zealand invention or not: was it invented *in* New Zealand or *by* a New Zealander? Was it something brand new; did it have some impact on the world; has it stood the test of time; and mostly – is it an interesting story?

Our greatest strength and our greatest weakness

We have a great deal of admiration for New Zealand's present and past inventors. In our many discussions with and research about inventors and inventions, we see a clear trend of talent, passion and in some cases – as they put it themselves – 'bloody-minded stupidity' that marks out the truly successful from those just dabbling.

New Zealand's remote geography is historically both a difficulty and a boon. The tyranny of distance creates the opportunity of independence. With no easy access to overseas technology, early New Zealanders had to innovate, finding solutions of our own to problems of our own, often able to think outside the box, simply because we were thousands and thousands of miles away from the box.

We call it No. 8 Wire thinking, academics call the concept 'bricolage', and 200 or so years of New Zealand's history has bred this into our culture as an attribute to be admired. This approach served us well for most of our history.

But not any more. Now it's clear this DIY thinking, this bricolage, has serious unintended consequences. While New Zealanders have the world's lowest 'power : distance ratio' (respect for authority) we also have one of the world's most individualistic cultures. A recent study of the psychological profile of New Zealand leaders showed we are very similar to the rest of the world in all dimensions of psychology – apart from two. We trail the world noticeably on our 'willingness to take on advice' and we exceed on the 'belief we can do anything ourselves'. This deeply ingrained personality, caused by – as Sir Peter Gluckman calls it – our 'frontier mentality', has got us to where we are now, but it is holding us back from getting much further.

There is another irony to the No. 8 Wire paradigm that many inventors have pointed out to us – while Kiwis have made heroes of the inventors of the past, the inventors of today find it hard to get taken seriously, to raise capital and to get help. Ian Taylor from Animation Research Ltd (see page 162) believes New Zealanders can do anything, if we'll let each other: 'I have often despaired that in New Zealand we are very quick to talk up the innovative abilities of Kiwis, our can-do attitude, but when it comes to trusting in those same people to deliver, the walk doesn't often match the talk.'

The complexity of the new world – 'everything easy has been done' – and the interconnectedness of everything; the ubiquity of information since the internet revolution; the rise in importance of non-nation states in the world (multinationals, for example); the exponentially increasing sophistication of consumers and markets – all this means that collaboration, learning from others, sharing information, embracing specialists and being prepared to fail fast are essential attributes for the current global reality. These are all things that are almost against our very nature. Kiwis need to get re-wired.

If you don't believe us, believe the numbers. According to the Organisation for Economic Co-operation and Development (OECD) Statistical Database, from being a world leader in the past, we were 22nd in 2008 in terms of patents filed to cover an invention in the United States, Europe and Japan. We had 3.3 'triadic' patents per million people, which is 20 times less than Switzerland, which had 66 per million, 10 times less than Finland and three times less than Ireland.

Why, when we think of ourselves as inventive, do we file so few patents? In the modern world, invention comes from research. In 2009 we spent 1.3 per cent of gross domestic product on research and development. This is half of the OECD average and three times less than Finland, a country of very similar size and population to New Zealand. It's worse when we look at what our businesses spend on R&D, which is one third of the OECD average. For every dollar spent on research by our primary sector (dairy, farming, forestry, fishing, etc.) the government chips in two dollars of R&D. For every dollar spent on R&D by our manufacturing sector, the government chips in 30 cents. For every R&D dollar invested by our services sector, the government contributes almost nothing.

Economists have shown that innovation and productivity go hand-in-hand. Which explains why our prosperity lags behind other countries of our size in the OECD.

It's a problem. It's a government problem, but it's also a company problem. And underlying all of that, it's a cultural problem. If we want books like this to be shorter in future, the sad fact is we're going the right way about it.

No. 8 Re-wired: our hopeful future

What might a re-wired New Zealand look like? Here are some characteristics which make up the No. 8 Wire paradigm and how they might better serve us.

FROM	TO
Alone in a shed	Collaborating with experts
Making a solution to solve my own problem	Understanding the customer's problem
Making do with what's at hand	Using the best possible technology from around the world
Distrusting others	Sharing information with others and mutually benefiting
Doing it all myself	Working with specialists in design, marketing, production, etc.
Holding on to the invention	Letting the invention go out to be criticised, commented on and added to by others
Making it 'good enough'	Perfecting the design and functionality
Standing apart from the world	Being open to global cooperation

We don't advocate throwing the baby out with the bath water. There are attributes of the No. 8 Wire mentality and the New Zealand psyche that will continue to serve us well. Our willingness to challenge the status quo, our use of unexpected tools to solve problems unconventionally, and the good nature with which we accept challenges and failures are all attributes we must retain.

We also have leaders and role-models in our midst. In February 2014, a single episode of *Top Gear* aired in Britain that featured not one, but two breakthrough pieces of New Zealand technology. First Jeremy Clarkson raced the newly released Gibbs Quadski, the world's first commercialised amphibious vehicle that drives fast on land and water (see page 110). Then he track-tested the brand new McLaren P1 – the most advanced production car the world has ever seen (see page 225). He didn't mention New Zealand, because, though the names written on those vehicles were Kiwi ones, the production of both has long been offshore. We could choose to lament that, or we could choose to see these as examples of products that have successfully been taken global.

Over in leafy Parnell in Auckland, biotechnology company Lanzatech (see page 178) gave us some clues about what the future might look like. They assembled engineers and scientists from many nations, with global investment, diverse ownership and customers in China, to create technology that is leading the world. They don't see barriers to doing their world-leading research in New Zealand; they have worked out how to turn our relative isolation, our culture, our land and environment into an asset that allows them to recruit talent from around the world. They know that New Zealand's rich cultural history has made us determined and resourceful – attributes they admire. They are adding to that history by demonstrating global collaboration and, as Sir Paul Callaghan put it, creating the place where talent wants to live. Meanwhile, nearby in Auckland, innovative power company PowerbyProxi (page 216) demonstrated that kiwis can collaborate, and can commercialise IP, when they leveraged inductive power technology from the team at the University of Auckland (page 214) that is in itself world-leading. They no longer see themselves as a Kiwi company going global; rather, a global company based in New Zealand. This company is not only re-wired, they dream of no wires.

All this – the urgent call to action, and even more, the exemplars of success – matters greatly. Because productivity is closely tied to innovation, our economic security depends on us finding a way to become once more the inventive nation that we are so proud of being. We hope the old and new inventions in this book might help inspire New Zealand to move from No. 8 Wire to No. 8 Re-wired.

If you have comments, or suggestions about the contents of *No. 8 Re-wired*, or you want more information, please visit www.no8rewired.co.nz. But for now, read on and be as surprised and inspired as we were by what some crazy Kiwi had created out of just a piece of No. 8 Wire – and imagine what we can do when we truly become No. 8 Re-wired.

JON BRIDGES & DAVID DOWNS
June 2014

1. Cheating Gravity

THE JETPACK, BUNGY JUMPING AND THE FIRST (US) SATELLITE

For centuries, even before Newton gave it a name, mankind has been seeking a way to break free from the shackles of gravity. With Richard Pearse (see page 78) Kiwis were among the first to create powered flying machines, and you could say that Hillary's ascent of Everest in 1953 was an attempt to cheat Earth's gravity by getting closer to the heavens.

In this chapter we'll see that our ambition reaches even further than that – out of Earth's atmosphere and into space itself.

Some of the reasons to escape gravity's relentless pull were out of necessity (creating more fertile farmland), some were more frivolous (bungy jumping, perhaps even kites), but what's surprising is how our dedication to the war on gravity is delivering far above what might be expected for a small country like ours.

The Martin Jetpack

GLENN MARTIN FINALLY BRINGS US THE FUTURE

144

The Martin Jetpack weighs 144kg and uses a 2-litre V4 petrol engine.

Right: ZK–JME, also known as The Martin Jetpack.

Since science fiction invented the future in the 1920s, there have been a few inventions humans hungered for. Disappointingly to all of us, almost none of them have materialised, even though 2014 is way beyond the promised delivery date. We don't live in pods under the sea, we don't instantly travel to other times or planets, and our dinner does not come in pill form. But the future is finally on its way, and it's a New Zealander who is bringing it to us – in his jetpack.

The history of jetpacks is long, and hundreds of prototypes have been made all over the world. The most famous of them is the Bell Rocket Belt, powered by tanks of expensive hydrogen peroxide which decomposes into steam and oxygen, blasting out of nozzles on the pilot's back and thrusting them into the air.

Before you knew it there were rocket packs popping up everywhere. At the opening ceremony of the Los Angeles Olympics – in the 007 movie *Thunderball*, even for the audiences at Michael Jackson's Dangerous World Tour – but there was a major problem.

The jetpack would only work for 30 seconds at a time, it screamed like a banshee and the pilot had to weigh less than 60kg and wear asbestos pants because the exhaust came blasting out at 740°C. In the end even the military gave up on jetpack research, putting it into the basket of excess difficulty. It looked like the jetpack would end up a novelty footnote in aviation history. The future would have to wait.

Enter Glenn Martin. And when I say enter I mean he entered his garage. Martin was studying biochemistry at the University of Otago when a simple question from his mates at the Captain Cook Tavern – 'Why are we not yet flying to work with jetpacks?' – sparked a lifelong quest. At the expense of his studies he began to learn about the mathematics of flight and tinker with the jetpack idea in his garage.

Martin was used to tricky challenges. As a child, his father used to task him with 'intellectual exercises' where he would have to solve interesting theoretical problems. So he went to the Science Library the next day and, after reading about the Bell Rocket Belt, set himself the challenge of designing the world's first practical jetpack – one that could carry a 100kg person for 30 minutes using normal petrol. How hard could that be? That was in 1981.

In 1984, after three and a half years of calculations and research, where he would often learn the maths theory he needed by sneaking into maths lectures he wasn't enrolled for, Martin had a eureka moment. He realised that the jet engine was entirely wrong for the jetpack, but that a ducted fan would be perfect for the job. Doing the maths, he confirmed that his challenge was entirely possible with a ducted fan – theoretically. Now he couldn't give up.

To support his obsession, Martin and his family moved to Christchurch to be near the University of Canterbury College of Engineering. He would spend a few years working, then take a few years off to work on the jetpack, and then repeat the process. Over the next 30 years Glenn mortgaged his house three times. It was an arduous process, cloaked in strict secrecy, and rewards were few and far between. Many jetpack prototypes were designed, built, tested, then discarded – all on paper. The first test flight didn't come for 16 years – about 15 years after most people would have given up.

In 1997 the jetpack was ready for its first test flight. Martin chose his wife Vanessa as the test pilot: 'Her qualifications were that, unlike me, she was under 55kg and she could keep her mouth shut.' She and the jetpack were attached to a pole in the garage and the jetpack simply picked her up and put her down.

Martin can't even remember his own first flight exactly; the tests were so gradual. He spent hours wandering around with just enough power to keep him on tippy-toes to test the control system. Then he would be a centimetre off the ground, then half a metre. As confidence in the machine grew, the flights became longer and higher.

In 1998 Martin finally quit his job, started Martin Aircraft and became the full-time chief executive of an aviation company.

The first public demonstration of the Martin Jetpack was in 2008 at an air show in Oshkosh, Wisconsin, and for a while it looked like the unveiling of the jetpack was imminent. In 2010, *Time* magazine named the Martin Jetpack among their top 50 inventions of the year, and the company announced jetpacks were soon to be delivered.

In 2011 New Zealand went jetpack crazy. TVNZ's *Sunday* programme filmed a demonstration of the potential of the jetpack, which flew perfectly in control to 3km high, piloted by a dummy called George Jetson. Then, in a simulated engine failure, the ballistic parachute deployed perfectly and George and the jetpack floated safely back to Earth.

At the time of writing, the commercial release once more seems imminent. A brand new prototype has been made – the P12 – with the ducted fans repositioned from pilot's shoulder height to waist height, and new improvements are being made to the engine. The Martin Jetpack weighs 144kg and uses a 2-litre V4 petrol engine, which powers twin ducted fans pointing downwards for lift. Of course, it's Martin's own specially designed petrol engine, not just the motor out of your average Camry.

The P12 is designed for the needs of its first customers – search and rescue, paramedic, ambulance, fire and border control services. In August 2013 the Civil Aviation Authority gave approval for New Zealand testing of the P12 after looking at it long and hard, and scratching their heads over exactly how to class the aircraft. It isn't a plane and it isn't a helicopter. They ended up classing it as a microlight, and the prototype P12 is officially registered as ZK-JME.

Nobody is yet willing to put a date on when the P12 will be released for sale – Martin is hoping it'll be 2014 – but the company is busy developing the necessary training manuals and maintenance programmes for the product.

For his part, Martin is determined to be the first to fly the Martin Jetpack across Cook Strait and the English Channel, 'although I've had a few arguments with a certain guy called Branson about who'll be first to do that!'

Meanwhile the aviation world is watching and waiting with bated breath. When asked how he hasn't given up sometime between 1981 and now, Martin says it's the conviction that jetpack flight is possible – a conviction based on science and mathematics. That, and 'bloody-mindedness and stupidity'.

After 32 years of development, it must be odd for Glenn Martin that his jetpack won *Popular Science* magazine's 'Best of What's New' award in 2013. But that's the thing with the future – no matter how long it takes, it's always new when it gets here.

Pickering's Satellite

ACTUALLY, YOU DO HAVE TO BE A ROCKET SCIENTIST TO WORK HERE

106

Explorer 1 **orbited Earth every 106 minutes, powered by batteries that helped it transmit data back to Earth.**

Right: *Explorer 1* satellite team in 1958, holding aloft the satellite they designed. From left: William Hayward Pickering, James Alfred Van Allen, and Wernher von Braun.

In the mid to late 1950s, the space race was well and truly on. The US and the USSR were staring at each other across the Bering Strait and daring each other to outdo as they each got closer to the ultimate goal of putting a man in space.

Imagine the Americans' surprise and envy when, on 4 October 1957, the USSR launched *Sputnik 1*, the first man-made object into orbit. It orbited the Earth every 98 minutes, passing over the continental United States seven times a day. Amateur radio fans could tune in to its signals and hear the strange beeping emanating from the unseen object many miles above their heads. It was up there, taunting them.

There was no way the US of A could let that one go, so in November 1957 they gave the job of creating their own satellite to the Jet Propulsion Laboratory (JPL) of the California Institute of Technology. The JPL was headed by a Kiwi, Dr William Pickering (b. 1910, d. 2004). Pickering was born in Wellington, but he moved to America to study for his PhD in physics. After a stint as professor of electrical engineering he helped set up and then ran the JPL, which developed, among other things, rockets.

In 1957, the task before the team at the JPL was immense. They were to create a satellite that was technologically the equal of *Sputnik*, and get it into space as soon as possible. They accomplished this in three short months.

The satellite they put together was dubbed *Explorer 1*, and it weighed a mere 10kg. It carried with it a cosmic-ray detection package, an internal temperature sensor, three external temperature sensors, a nose-cone temperature sensor, a micrometeorite impact microphone, and a ring of micrometeorite erosion gauges. Which, I'm sure you'll agree, sounds adequate for the purposes. Crudely speaking, the team at the JPL strapped all of this to the top of a rocket, pointed it into the sky and, on 31 January 1958, the US caught up.

One of the experiments carried out by *Explorer* was searching for the existence of charged particles high in our atmosphere. The satellite found these particles in a layer, trapped by Earth's magnetic field – dubbed the Van Allen radiation belts after James Van Allen, Pickering's colleague who had designed the experiment. The discovery of the Van Allen belts by the *Explorer* series of satellites was considered to be one of the outstanding discoveries of the International Geophysical Year, an international scientific project that lasted from 1 July 1957 to 31 December 1958.

Explorer 1 orbited Earth every 106 minutes, powered by batteries which helped it transmit data back to Earth. The batteries gave out after a mere 31 days, leaving *Explorer* floating in space as a hunk of useless but memorable space junk.

Pickering went on to become an influential figure in the American space race as it progressed over the following years, and the very public success of *Explorer* gave him the credibility and freedom to be included in other projects. He worked on plans for near-Earth satellites, deep space missions and the development of manned space travel, among other things. Under his leadership until 1976, the JPL carried on creating satellites and rockets which helped the US to explore space – *Pioneer 4*, the *Mariner* flights to Venus and Mars and the unmanned lunar landings of 1966/67 were projects undertaken and led by Pickering.

Pickering went on to become an influential figure in the American space race as it progressed over the following years, and the very public success of *Explorer* gave him the credibility and freedom to be included in other projects.

Being a Kiwi – albeit one with American citizenship – Pickering was awarded an honorary knighthood for his achievements. He also won many accolades and awards and appeared on the cover of *Time* magazine twice. He passed away in 2004 at the age of 93. In 2009, a peak in the Kepler Range in Fiordland was named Mount Pickering in his honour.

The New Zealand Space Programme

SPACE FLYING KIWIS

In 2009 New Zealand's first space flight made big headlines. The rocket, called *Manu Karere*, meaning Bird Messenger, was blessed by iwi, fixed with a $6 part from a local engineering shop and then launched from the Coromandel coast. It burned for 20 seconds, flew at 5000km/h and reached over 100km above Earth. For a short time it was the highest thing in the Coromandel, which is saying something.

Rocket Lab are based in Auckland. CEO Peter Beck is an engineer with no university degree who has learned rocket science by doing it. At the age of 18, Beck left his home town of Invercargill to work for Fisher & Paykel in Dunedin. He set up Rocket

2012

In late 2012 they demonstrated a rocket to representatives of their US military clients.

Left: Peter Beck and his Atea-1 series rocket Manu Karere (both actual size). Below: The inner workings of Beck's *Instant Eyes* product.

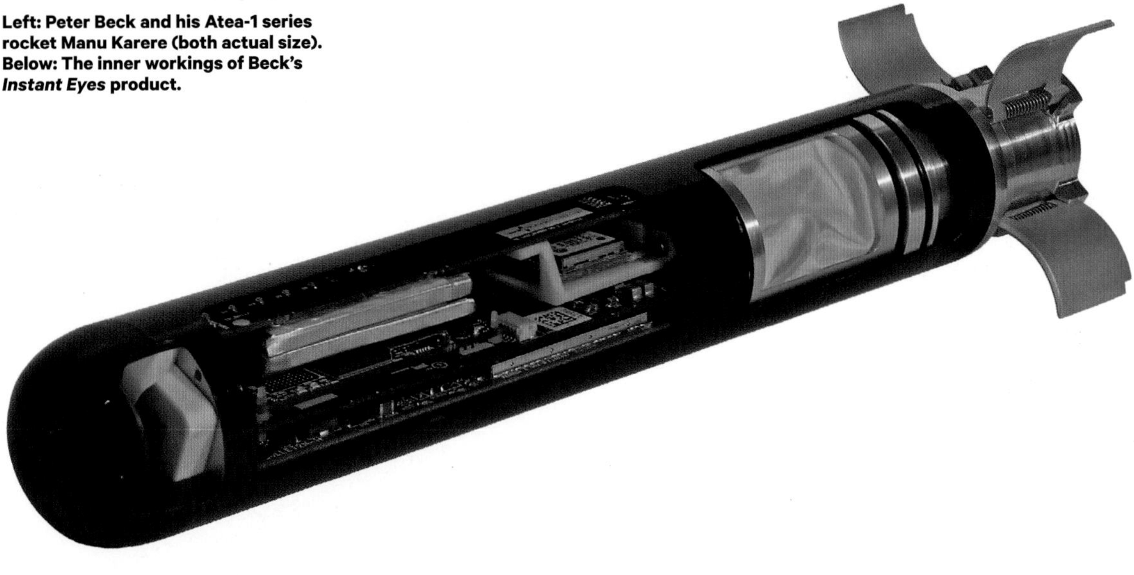

Lab in 2006 with funding from rocket-mad angel investor Mark Rocket (not the name he was born with). Now, at the age of 36, Beck already has a long history of innovation, multiple awards, and a good reputation in the aerospace world.

Since the first launch, Rocket Lab have begun to make world-first technologies. In late 2012 they demonstrated a rocket to representatives of their US military clients. The rocket runs on a viscous liquid monopropellant (VLM) fuel which is thixotropic – neither a solid nor a liquid. It has all the best properties of both of those sorts of fuel and, if successful, will be a major advance in rocket science.

They are also developing a new liquid engine called the Rutherford, intended for use in their first orbital rockets. *Popular Science* magazine gave Rocket Lab its 2011 'Best of What's New' award for *Instant Eyes*, an unmanned rocket that 'launches with the push of a button and snaps five-megapixel shots throughout the 120 seconds it takes to parachute 2500 feet back to Earth, transmitting them by encrypted Wi-Fi to the soldier's phone, tablet or laptop. Once the unmanned aerial vehicle (UAV) hits the ground, it self-destructs.'

Beck says that although New Zealanders are often bemused about a Kiwi space programme, it isn't hard to interest overseas companies in his developments.

Beck says that although New Zealanders are often bemused about a Kiwi space programme, it isn't hard to interest overseas companies in his developments. 'If you've got a technology that's superior, people will listen.'

Among Rocket Lab's clients are the Defense Advanced Research Projects Agency or DARPA – the US Department of Defense agency that gave the world the internet – and the US Office of Naval Research. Beck says Americans often introduce him by saying, 'This is Rocket Lab, the New Zealand space industry!' Then they laugh because one company is our whole industry. We laugh because one company is one more than we've ever had before.

The Springfree Trampoline

NO MORE PINCHED BUMS

Right: Engineer and father (and graffiti artist) Dr Keith Alexander.

Where's the most dangerous place a parent could imagine their kids to be – swimming in a pool of sharks, suspended above a volcano or on a trampoline in their own backyard? Thankfully University of Canterbury professor Dr Keith Alexander has made at least the latter safer by inventing the Springfree Trampoline.

In 1987 Alexander's wife wouldn't get a trampoline for their kids because of the danger of snapping limbs. Being a good father and a curious engineer he saw this as an engineering challenge, and set about improving the tramp. It turned out to be harder than expected, so he enlisted the help of his graduate students. Fifteen years later the Springfree Trampoline was a reality.

The final design uses fibreglass-reinforced plastic rods instead of those nasty old springs. The rods are placed diagonally from the metal frame and support the mat from beneath so that anyone bouncing on it can't land between the springs and hit the frame – no more chipped teeth! It's unclear whether this also solves the old 'static shock when dismounting on a hot day' problem. As well as safer, they're still just as bouncy – test trampolines are subjected to three million bounces, simulating 10 years of usage.

And good bouncing also makes for good business – the Springfree Trampoline has become a successful Kiwi invention and an internationally sound business proposition, selling in large volumes in Australia, Europe and the United States. Set up as a commercial venture with the help of the university's commercialising division, they manufacture a number of the components in New Zealand: the rods are extruded in Gisborne and sent to China for assembly. Demand for over 40,000 units a year has meant their Chinese manufacturing facility has had to open a second factory.

It took a lot of work and a few risks to make the trampoline a commercially sound proposition, including a substantial amount of personal investment: over $250,000 on intellectual property protection alone. Initially, no one wanted to support the idea and Alexander was forced to work with offshore investment until the university agreed to help back him.

Alexander still consults for the company he helped set up, but has kept his university day job. He has a few other ideas brewing where he sees an engineering mindset can be applied to a common problem. Who knows, next he may solve the issue of the killer swingball.

Kites and Saws

I'VE GOT THE WORLD ON A STRING

40

The world record holder for the largest kite ever flown – with a total area of 1019m², it's over 40m wide and weighs over 200kg.

Below: A giant Manta Ray kite.

Peter Lynn (b. 1946) is a man somewhat obsessed with kites – a strange occupation but as a world-renowned kite designer, Peter has managed to make a living dangling things from the end of a bit of string. It's the new New Zealand success story – pick a niche, and then be the best in that niche. Before he started with his kite obsession, Lynn also created a 'tipping blade' sawmill system in the 1970s. He realised there was a better way for sawmills to work than having two blades, and making a sawmill that could do the same job for the same energy with just one blade might be a winner. It was. He patented his idea, licensed it out and largely forgot about it, and the system is now adopted broadly in sawmill systems around the world. Moving on from sawmills, he became the holder of a number of patents around the world for innovations in kite design ('A wing for a traction kite comprises a plurality of cells formed by chord-wise-extending ribs . . .') and also the world record holder for the largest kite ever flown – with a total area of 1019m², it's over 40m wide, weighs over 200 kg and was in the shape of a giant Kuwaiti flag. Lynn lives in Ashburton but spends much of the year travelling, and is a pioneer in the use of kites for sports – he developed the sport of kite buggying; and his kiteboards and kitesurfers can often be seen in the windy, low-tide areas around New Zealand and the world, terrorising the locals.

Top-dressing

CLOUDY WITH A CHANCE OF FERTILISER

For most of our history, New Zealand has basically been a big farm, pumping out products for the United Kingdom. It's a gross over-simplification, but essentially the colony of New Zealand, from its establishment in the early 1800s through to when the UK joined the European Union in the 1970s, was the place that grew the Mother Country's meat, wool and milk – all from the 40-odd million acres of farmland we are blessed with.

As any farmer will tell you, the constant sowing, growing and reaping of crops will deplete the natural minerals in the soil over time. Leaving aside the wider ecological implications of felling forests to create pastureland, the fact that pastureland becomes less productive is a real challenge to efficiency. Add to that the fact that sowing seeds, particularly into the hills and vales of steep New Zealand, is difficult, expensive and inefficient: now you've got a problem that needs ingenuity and inventiveness.

In 1906, one John Clervaux Chaytor (b. 1836, d. 1920) became the first person in the world to apply agricultural materials aerially – he went up

2

He experimented with sowing lupin seeds by sewing a sack of seeds to a downpipe, then flying at different heights to work out the right dispersal rate.

Above: That plane is not in trouble, it's spreading fertiliser in inaccessible hill country.

1939

Throughout the period 1939 to 1943 he experimented until he had some quite specific and exact results to show that this method of dispersing seeds and fertiliser was extremely economical.

Left: A modified World War II Grumman Avenger spreading superphosphate fertiliser at Ohakea, circa 1948.

It was the perfect time to start the endeavour – after World War II ended there were a lot of qualified pilots looking for jobs and a number of surplus planes left over from the war efforts.

in his hot-air balloon and threw seed into the air so that it would spread over the valley below. The location of this feat is often incorrectly quoted as Wairoa: it was in fact in Wairau, in Marlborough, on the family farm 'Marshlands'. This mechanism of spreading seed was one the family kept at, and as a young man John's son Edward Chaytor (a famous soldier during World War I, later knighted) spread grass seed over the family farm from a hot-air balloon.

Most historians credit Alan Pritchard with pioneering what we now call 'aerial top-dressing' – flying a plane over a patch of ground and releasing fertiliser or seeds (actually often both at the same time) over the land. Pritchard was a pilot with the government's Public Works Department in the 1930s and 1940s. He thought of the idea of sowing seeds from his plane after eating grapes while flying and throwing the seeds out of the window. A few days later he experimented with sowing lupin seeds by sewing a sack of seeds to a downpipe, then flying at different heights to work out the right dispersal rate. Strictly speaking, his employers didn't really know about his aerial experiments, and he forged the logbooks of his flights to allow him more time to trial his ideas.

Throughout the period 1939 to 1943 he experimented until he had some quite specific and exact results to show that this method of dispersing seeds and fertiliser was extremely economical. After he published his results, a government minister he regularly piloted asked him how he'd worked it all out. When Pritchard admitted what he'd been doing with the logbooks, the minister gruffly congratulated him, and told him that if anyone had an issue with it, to send them to him. With that endorsement, top-dressing was ready for mainstream use.

It was the perfect time to start the endeavour – after World War II ended there were a lot of qualified pilots looking for jobs, and a number of surplus planes left over from the war effort. Indeed, at one point it was even suggested that the RNZAF should take on top-dressing as a core function, but despite undertaking some successful trials using DC3s and other air force planes at Ohakea, the private sector took on the task and a successful new approach to farming took off.

Known more commonly as 'crop dusting' in other parts of the world, top-dressing is still a core part of how we maintain the land. The lush pasturelands of New Zealand – and a number of happily employed private pilots – owe their success to Pritchard and his predecessor Chaytor.

Cablecam

OSCAR-WINNING INNOVATION
IN THE FILM INDUSTRY

1993

He spent three years tinkering with the system and using tricks from his fishing and timber industry experience before getting a patent in 1993.

Right: Trou Bayliss' new camera system 'Flite Line' in use for a State Insurance ad in 2012.

This may well be the only invention in the book to be able to say those familiar words 'I'd like to thank the Academy . . .' – for the Cablecam, and its inventor Thornton Bayliss (or Trou as he likes to be called), won an Academy Award in 1998. It was a pretty big deal for a guy from Te Puke who left school at 15 to surf, then got a job working the camera on a small film shoot and was hooked – eventually ending up in Hollywood working alongside some of the big names in the industry.

The idea behind the Cablecam came when Bayliss realised there was a need for a system that film cameras could use to get into tricky-to-film places, where a traditional camera couldn't go and where a helicopter or other flying device wouldn't work. He spent three years tinkering with the system and using tricks from his fishing and timber industry experience before getting a patent in 1993.

The Cablecam was initially designed to be used for sports, and one of its first uses was filming the Kentucky Derby horse race. It was then used for the Winter Olympics in Norway in 1994 before crossing over to the film industry proper. It's been used by some big-name directors like Spielberg, Cameron and Burton (Tim, not Richard). Today hardly an hour goes by when the system is not being used, particularly in televised sport.

The Cablecam system works by rigging cables between frames, and in the original version they suspended a platform device underneath for an operator to sit or lie on. Unmanned cablecams are now more common, but imagine the thrill of travelling up to 130km/h on a thin cable while peering down a lens at the action below.

The system is well known for its smooth operation and easily repeatable way of getting long tricky shots where other systems fail. It can extend over 600m and – as they said at the Academy Awards – 'it can function at speeds and through perils that would be unsafe for on-board operators' such as down rivers or up mountains. It was used in the film *Dante's Peak* to do just that.

Speaking of that Oscar, Bayliss received a Scientific and Technical Award for the system, although he didn't believe it when he got the fax telling him he'd won – he thought it was a joke. When the invitation arrived with gold lettering on it and 'probably costing more than my couch', he realised it was legit, and picked up the award on the red carpet.

Bayliss continues to innovate with camera and film systems: along with Kiwi colleague Harry Harrison he developed the 'Flite Line' camera wire system, which was used in a 200m single-shot ad for State Insurance in 2012. His new system is about a tenth of the size and weight of the original Cablecam, it's self-propelled, it can go up to 190km/h on an 800m long 'string' (as Bayliss calls it), and 'it's great for shooting noisy sh*t that goes fast!'.

Ski Plane

THAT SNOWPLANE,
THAT'S MY INNOVATION

Right: Harry Wigley (far left) with the first ski plane (an Antarctic Beaver) before take-off. Edmund Hillary is the one in the patterned jersey.

Henry Rodolph Wigley (b. 1913, d. 1980) continued the Kiwi tradition of being first with an aviation exploit and then having the world largely ignore it. In Wigley's case, he was the first to land on snow in a plane that had skis as well as wheels. He also survived, which makes the story even more interesting.

Wigley was born in Christchurch. His father started up what was to become Mount Cook Airlines and his mother designed their distinctive logo, based on the Mt Cook lily, which isn't a lily at all but a huge buttercup that grows in the region.

The young Wigley learned to fly at an early age, but in 1936 was lucky to survive an incident when the aeronautic plane he was passenger in had a near miss. The air force plane was attempting inverted flying, looping and rolling at low altitude and managed to clip a parked bus, losing part of its undercarriage as a result. The plane landed safely with no injuries to Wigley or the pilot, although two newspaper reporters in the bus did get a nasty fright.

The incident clearly didn't faze Wigley too badly, and he went on to be a World War II fighter pilot in the Pacific, a mountain climber, a national downhill skiing champion and an astute businessman.

On 22 September 1955, he attempted his own aviation feat of daring. With one Alan McWhirter as his passenger, he took off from Mt Cook village in a plane with a hand-crank attachment of his own invention with which he could lower either wheels or skis depending on whether he was landing on ground or snow. He had to lean out the window to operate it, but having spent years developing the system, Wigley was in no doubt that it would work, and he safely guided his four-seater Auster aircraft in to land on the Tasman Glacier.

He took off from Mt Cook village in a plane with a hand-crank attachment of his own invention with which he could lower either wheels or skis depending on whether he was landing on ground or snow.

This was the first time a plane had taken off on wheels, and landed on skis – overseas, aircraft were equipped with skis for snow take-offs and landings, but the use of both wheels and skis was unknown. This flight was repeated later in the day with Sir Edmund Hillary on board and again, the ski system worked flawlessly. So good was the system, that within a year Mount Cook Air Services Ltd was running commercial ski plane operations.

Henry (or Harry as he liked to be called), was knighted and went on to follow in his father's footsteps and become chairman of Mount Cook Airlines. In 1975 he recreated his original landing in the same aircraft, and with the same passenger as on the original flight.

Bungy Jumping

NEAR-DEATH PLUMMETING MADE FUN?

3.5

In 2012 Hackett and van Asch's company passed the 25-year milestone, with over 3.5 million people paying $180 each to cheat death and get a DVD of it.

Right: It seems that jumping with arms spread wide is more dramatic than having hands nonchalantly in your pockets.

I'll come right out and admit that I have never done it, and further, I am unlikely to ever do it. The idea of hopping out to the end of a wooden platform with my ankles tied together, tethered to the Earth by a long piece of what is essentially elastic, is both frightening and puzzling. Yet people by the hundreds of thousands – granted, many of them foreigners with little command of English, often drunk and far from the safety of their homes, possibly misconstruing the whole situation and thinking it's some kind of compulsory initiation ceremony or the only way to get a T-shirt in New Zealand – have paid good money to do it.

The idea of jumping off things is not a new one, nor is the idea of stopping oneself before one gets hurt. Again, like many inventions, it's not necessarily about coming up with a brand new idea; it can be about taking an existing one and making it work. The natives of Vanuatu have for centuries been

jumping off large towers with vines tied around their legs to break their fall. And their legs.

The English apparently had a try of it in the 1980s, but it didn't last. What this crazy idea needed was a plan to make it less, well, fatal seeming. Alan John (AJ) Hackett (b. 1958) had just such a plan.

Hackett was a speed skier from New Zealand who heard of the strange Vanuatuan practice, and saw in it the germ of an idea. With colleague and fellow skier Henry van Asch, he developed the 'bungy' – a series of hundreds of strands of latex rubber, bound together into a cord. The bungy was extensively tested until they felt confident it would do what they wanted. After testing it from a number of bridges across the North Island, Hackett undertook his significant 'leap of faith' in Tignes, France, jumping in extreme wind and -20°C temperatures. He knew then that bungy jumping was ready for the mainstream – and set about making it a worldwide phenomenon.

One of Hackett's strong skills, apart from an apparent disregard for his own life, was as a showman. He's set a number of Guinness world records, and in June 1987 he ensured that the legend of the 'crazy Kiwi' would make bungy jumping a financial success by undertaking a bold move. In the middle of the day, he and a support person snuck up the Eiffel Tower in Paris and attached a bungy to the edge. Hackett then jumped off, plummeting down between the tower's wide steel legs. The deed earned Hackett an arrest by the French gendarmes, but also worldwide publicity, and the respect of thrill seekers everywhere and the French themselves. Indeed, at the top of the Eiffel Tower today there is a large memorial to Hackett's feat. 'Memorial' is not exactly the right word, but it has a nice ring of mortality that bungy jumping uses as one of its drawcards.

Hackett started the world's first commercial bungy operation from the Kawarau Bridge over the Shotover River near Queenstown in 1988. It should be pointed out that the safety standards of bungy jumping are extremely high, and this is thanks in no small part to Hackett himself, who pioneered a set of stringent standards (such as 'the bungy cord must be shorter than the height of the platform') to ensure that bungy jumping is as safe as any other sport where you throw yourself off something solid and into thin air.

Hackett has also earned a reputation as a shrewd businessman, and has managed to grow what was initially a 'crazy idea' into a global empire. In 2012 Hackett and van Asch's company passed the 25-year milestone, with over 3.5 million people paying $180 each to cheat death and get a DVD of it. There are now full-time operations in Australia, France, Canada, Macau, Germany and, of course, New Zealand, and a significant set of thrill rides in Russia. But I'm still not going to do it.

2.

All Creatures Great and Small

SHEEP, EARTAGS AND THE TRANQUILLISER GUN

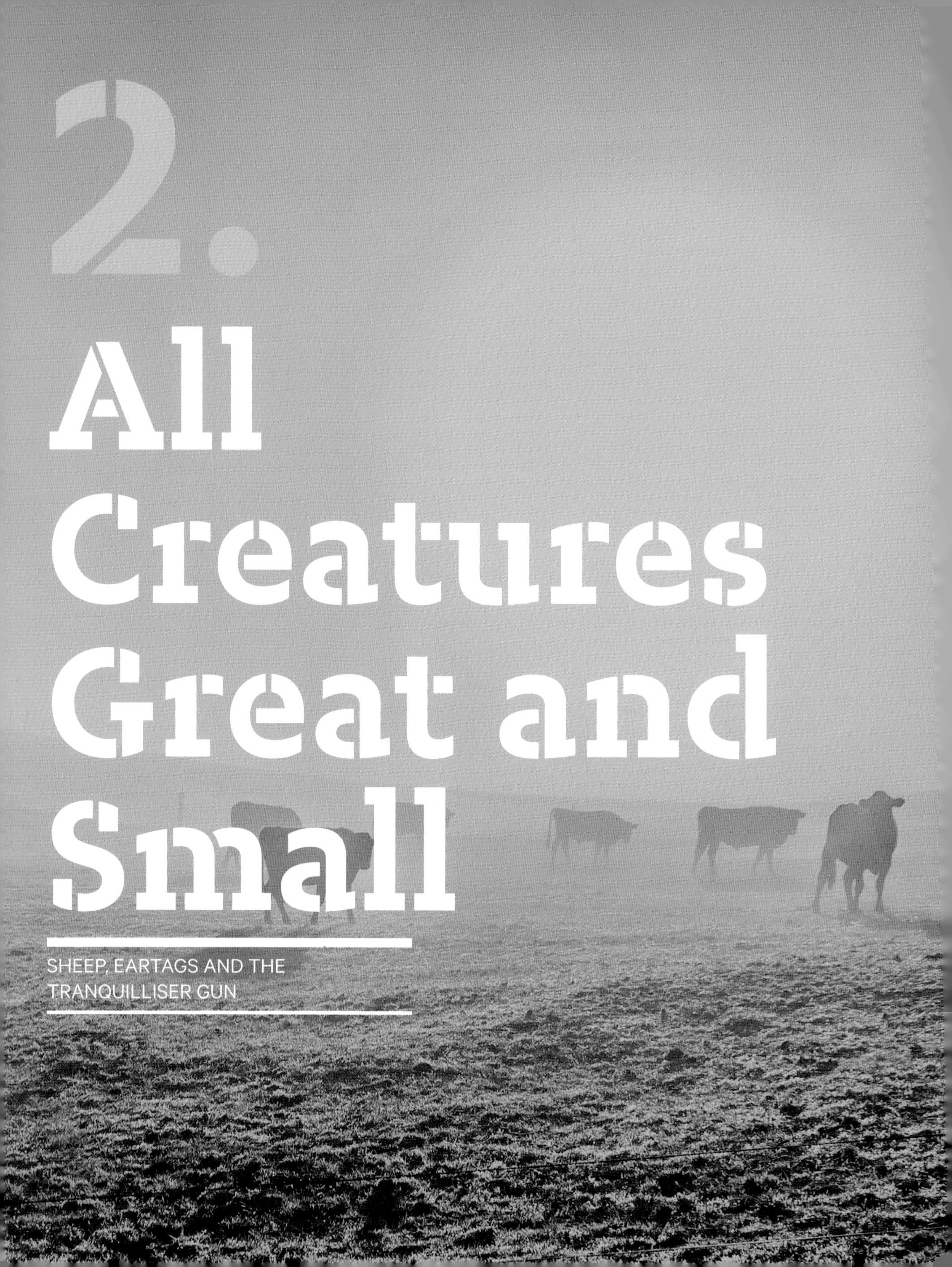

New Zealand had, and to a large extent still has, an agrarian economy. In all our traditions and popular lore, the country was built on the back of good honest God-fearing farmers, quietly tilling the soil and tending the flocks, and in the process ensuring our country's place as the 'farmyard of England'.

It's only natural, then, that many New Zealand inventions and innovations have a 'farming' air about them. Not (just) an odour of silage – more the impression that these inventions were born of Mother Necessity, and sired by Father Practicality. It is in all of these agricultural inventions that the No. 8 Wire tradition shines the brightest.

Because New Zealand farmers really did find themselves isolated – they were far from Mother England, far from authority, far from help, often just far from the shops – and working in a new land that posed new challenges, where the accepted way of doing things often didn't work any more, and didn't seem to hold as much authority anyway. All this novelty, distance and the removal of the normal rules led to great innovation, and oftentimes this innovation went on to change the world.

The Milk Production Meter

THE BIRTH OF TRU-TEST

1963

Prior to Hartstone's invention in 1963, farmers and dairy companies couldn't accurately measure the output of an individual cow as it was being milked.

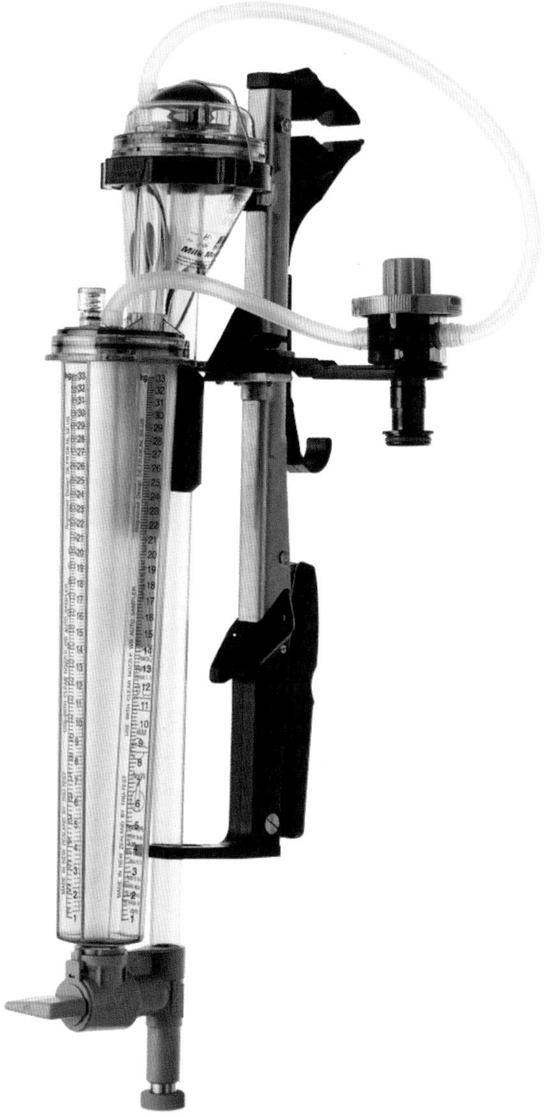

Right: The Tru-Test WB AutoSampler is a recent model used for large herds.

No doubt about it, John Hartstone (b. 1923) is the quintessential 'bloke in a shed' inventor. Not only did he dream up his invention in the middle of the night, but he also yelled 'Eureka!' when it worked. Hartstone turned that eureka moment into a globally successful product with 95 per cent market share and tens of millions of dollars of sales. He also significantly contributed to the improved productivity of dairy farmers the world over. So what was the great idea? A meter for measuring the milk from an individual cow.

Prior to Hartstone's invention in 1963, farmers and dairy companies couldn't accurately measure the output of an individual cow as it was being milked. The British Milk Marketing Board lamented the lack of an accurate measure in a magazine article Hartstone was idly reading during a night of insomnia. He thought about the issue and realised he might

24

From an idea in the middle of the night to a working prototype in less than 24 hours!

Left: A brochure for the 1970s model. You can tell because it's got large sideburns.

The reason it is so important to accurately measure the milk coming from an individual cow is that the information is key to breeding programmes.

have an idea of how to solve it. He knew a lot about cows: he milked 130 of them twice daily. He wrote down all the principles he thought would be important for the measurement system, and as he did so, he recalls having a 'blackboard in my mind' upon which he could see the solution developing. By 5am he was pretty sure he'd cracked it. He trundled off to do the morning milking, and straight after that he hit the nearby big town of Otorohanga to pick up parts for a prototype.

He went to see Tom the plumber to get copper piping and a ballcock, and Harry the chemist for a glass measuring cup; and then went to have a chat with Ray, the local engineer and gunsmith. He brought an empty lemonade bottle he had knocking around the farm and put the whole lot together in time for the second milking of the day, when he tested it on the spot. From an idea in the middle of the night to a working prototype in less than 24 hours! It worked perfectly and became one of New Zealand's most successful gifts to the farming world. The Tru-Test milk meter was born.

The reason it is so important to accurately measure the milk coming from an individual cow is that the information is key to breeding programmes. Milk production is hereditary, so if Daisy gives below average milk volumes, then it's likely Daisy Jnr will too. Much better to know that, and not to breed from Daisy, but instead from Clarabelle, whose milk production is above average. Do that enough times and you can significantly increase the output from your herd. Indeed, over the past 50 years, the average output per hectare of land has nearly tripled – and a sizable portion of that increase is because the milk meter allowed for more accurate herd assessment and selective breeding. Our cows are now so much milkier!

Hartstone spent a few years tinkering with his design and working out how to mass produce it, then he and three of his sons became mobile evangelists for the product. They travelled New Zealand and the world convincing authorities to approve the device, and then persuading farmers to use it. They were very successful, and Tru-Test has gone from strength to strength.

Hartstone retired in 1987, but the product he dreamed up has stayed largely the same, and today has a majority of the world market, with an estimated 500,000 meters in use every day in over 100 countries. This 'bloke in the shed' had a great idea – but more than that, he worked out how to turn it into a truly global winner.

Home-made Yogurt

EASIYO IS THE FIRST IN THE WORLD

If all goes to plan for an inventor, and he or she has all the success they could hope for, then they'll go through what we might call the 'inventor's life cycle' – idea leads to invention, prototype works, goes through development to market launch, sales success gives rise to the sale of the company, retirement in luxury, and ends up as a section in a bestselling book. Of course every inventor is different. Some would-be inventors never even have the idea; some never retire. This is the story of an inventor who got everything just right, then came back for more.

38

With most home-made yogurt kits you have to keep the yogurt at 38°C for seven hours, which, unless you live in Darwin, is tricky.

Right: Len Light was an engineer and science teacher at Westlake Boys High when he and his wife Kathy launched EasiYo.

We've all seen EasiYo in the supermarkets, and many of us have an EasiYo maker in the pantry. EasiYo's inventor, Len Light, was the first in the world to invent a successful home yogurt-making kit. The need was there: cheap supermarket yogurt often contains preservatives and other nasties. Expensive yogurt is…expensive, especially when you've got eight children as Len did. And *all* storebought yogurt suffers from the fact that the precious bacterial culture – the part that brings all the health benefits – deteriorates over time. By the time you get yogurt home it isn't nearly as good for you as it was when fresh. If only everyone, everywhere could easily eat cheap, fresh, pure yogurt…

With that idea in mind Len Light did what any Kiwi man would do. He disappeared into his shed. The 'Easi' part was the hard part. People have always made yogurt at home, but only dedicated people who are probably, let's face it, slightly nutty. You need a double-boiler and a 'culture' that's been kept alive, and you have to keep the yogurt at 38°C for seven hours, which, unless you live in Darwin, is tricky. In other words making yogurt at home was hard and nobody really bothered to do it.

Light successfully worked out how to dry not only the milk, but also the lactic cultures. When you add water and put the litre container of mixture inside the EasiYo 'thermos' of boiling water overnight, the freeze-dried cultures wake up and start fermenting the milk, turning it to yogurt. Eight hours in Light's special yogurt incubator and you've got your own fresh yogurt in the kitchen. No preservatives, and teeming with *Lactobacillus bulgaricus*, *Streptococcus thermophilus* and *Lactobacillus acidophilus* (all good things, promise).

In 1992 Push Push was number one on the pop charts and *Shortland Street* premiered on TV2. That excitement was only increased when Light launched EasiYo at the 1992 Auckland Home Show. EasiYo soon became a household name. Westland Milk Products bought part of the company in 1997 and the rest of it in 2010. It's now the number one make-at-home yogurt brand in the world.

Having sold EasiYo didn't mean it was time for Light to rest. He jumped straight back into setting up another business. Then again, when you've got eight children, rest is probably never an option.

Sonic Electrospinning

SOUNDS LIKE A NEW DANCE CRAZE

Aptly named company Revolution Fibres, working out of Henderson in Auckland, have exploded on the scene with a burst of energy and an incredible power of innovation. Their invention is Sonic Electrospinning Technology – a new way of spinning nanofibres.

Nanofibres are fibres 5000 times thinner than a human hair. They can be spun from organic proteins – Revolution Fibres use collagen derived from waste fish skins – and the gossamer fabrics they make are the subject of intense global research. Thousands of scientists around the world are coming up with ways to use them: human tissue engineering, textiles with special properties, protective materials, batteries and solar cells, acoustics, composite reinforcement for super-light construction, filters and absorbent materials.

Electrospinning around the world is very much on the lab scale, using needles and producing a fraction of a gram per hour, which is no good for commercial production. Revolution Fibres have invented a needle-less system. Sonic Electrospinning zaps collagen fluid with 40,000 volts, causing the droplets to stretch as it is spun out at 80m per second to form extremely long threads which are deposited to form a fabric. With new sonic looms, Revolution Fibres are aiming to produce hundreds of square metres per hour of this precious stuff.

Already they are producing nanofibre air filters, a nanofibre laminate called Xantu.Layr which can double the strength of things made of carbon fibre, and a skincare product called actiVLayr in which they infuse plant extracts into collagen fibre.

We're only at 'Once upon a time' in the nanofibre story, and there's plenty of time for Revolution Fibres to turn out to be the hero.

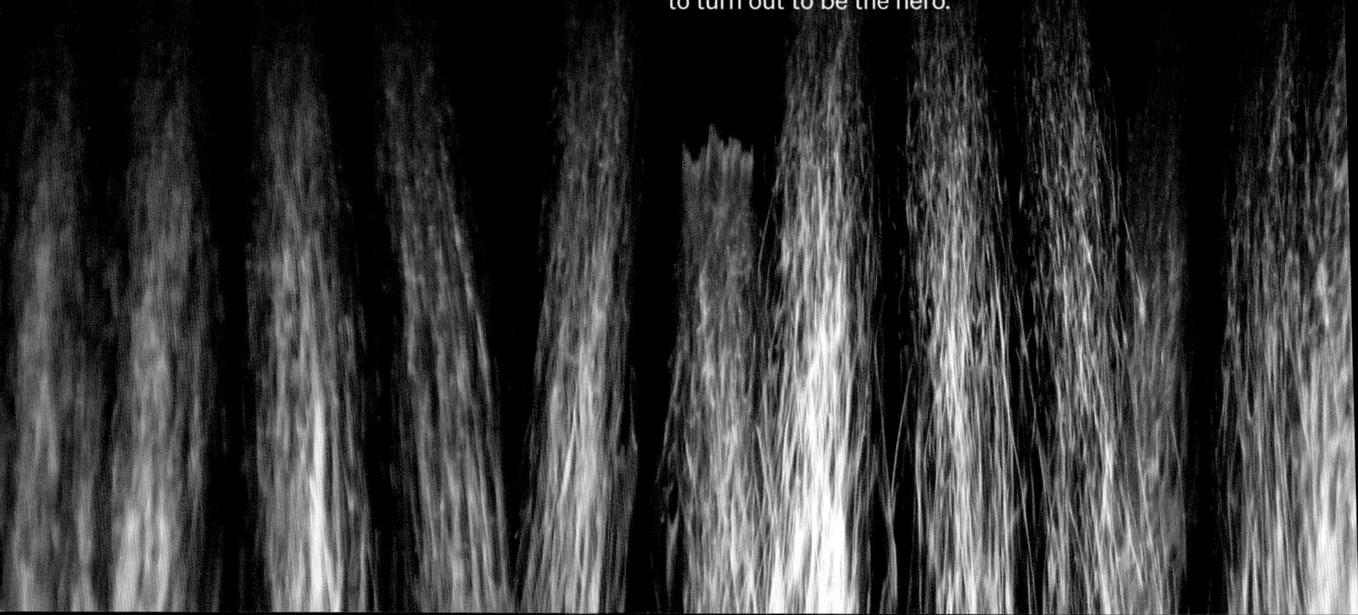

Below: Nanofibres extracted from fish collagen.

Woollen Shoes

DEATH KNELL FOR THE SOCK

If there's a product in the world, at some stage a New Zealander will have tried to make it out of wool. It's our national material. So far this has been successful with jerseys, face masks and lately even suitcases – and unsuccessful with hamburgers. You might be surprised that until 2013 nobody in the world had ever successfully made shoes entirely out of wool. Despite wool's wonderful qualities, a woollen fabric strong enough for shoes had never been invented.

Three Over Seven of Wellington has invented the world's first woollen running shoes and casual shoes. Founders Tim Brown and Michael Wilson (both former All Whites) and Tim's brother Paul were convinced that shoes could be made from wool – if only a strong enough fabric could be made. Reading an *Idealog* magazine they heard about research and development company AgResearch, so they picked up the phone. 'A world-renowned textile engineer answered his own phone,' says Brown. It turned out AgResearch had already been developing strong wool fabrics. A partnership was born and over the next couple of years the world's first woollen shoe fabric was developed and a patent applied for. The fabric uses a synthetic architecture to help provide strength, but puts wool next to your skin.

In 2013 Three Over Seven raised over $100,000 on crowdfunding platform Kickstarter to get the shoes into production. They also scored $200,000 from government-funded Wool Industry Research Limited, and money and mentoring from a British business incubation scheme. People are genuinely excited about this technology!

Wool wicks away moisture, is antibacterial so doesn't smell, regulates temperature well, and is comfy and sustainable. Wool Runners are machine-washable. They have just one seam in the upper and are so soft and comfortable they are designed to be worn without socks. Do you want some? Well join the queue.

Three Over Seven say the Wool Runner shoes will be in production soon. But they are really just a proof of concept for the breakthrough wool shoe fabric, which they hope to license to manufacturers all over the world. Then they can sit back and put their woolly feet up.

Milking Sheds

ROTARY AND HERRINGBONE SHEDS

70

By the mid 1960s, 70 per cent of new parlours were herringbone style – and not just here, but in Australia and the UK and all over the dairying world.

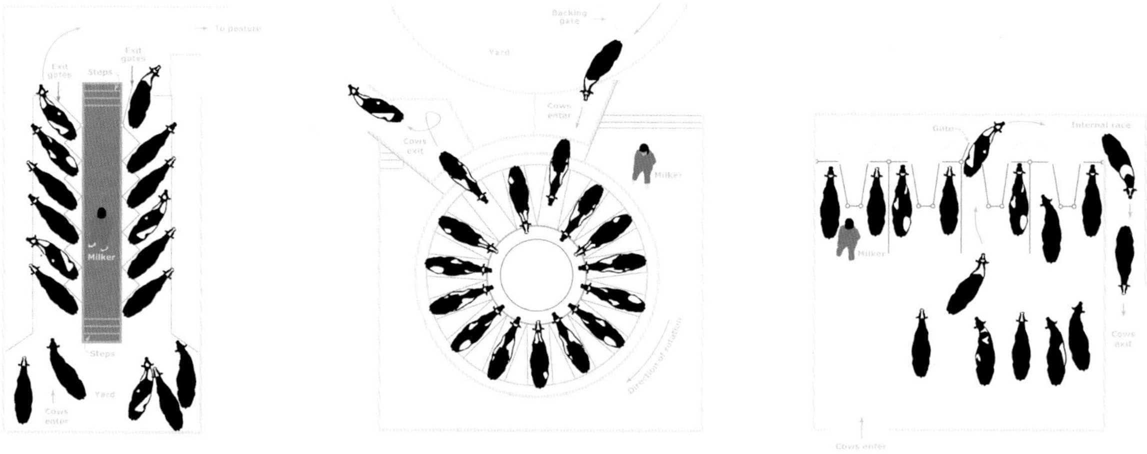

Left to right: herringbone, rotary and walk-through milking sheds.

Postwar prosperity in the New Zealand dairy industry meant a lot of farmers here had cash to invest in new machinery. In an era when the motorcar was becoming more common, and households were becoming mechanised with washing machines and fridges, farmers were actively looking for the latest technology to save them time and money. As a result, New Zealand's dairy industry became the most efficient and modern in the world. This hunger for mechanisation meant that cowsheds, toolsheds, woolsheds and every rural corner of New Zealand were glowing and buzzing with inventive individuals who would change the worldwide dairy industry forever.

The government had its hand in this push for mechanisation as well. With the 1952 Dairy Industry Act, the Department of Agriculture demanded higher quality and more hygienic dairy products and processes, and strongly encouraged Kiwi farmers to upgrade their milking premises.

The design of the dairy shed, or parlour, was the most basic and crucial thing to get right. The evolution of dairy practice had seen the simple 'back-out' shed give way to the 'walk-through' shed in the early part of the twentieth century. Each milker could milk around 42 cows per hour in a walk-through parlour, but it was back-breaking work bending to attach the machines to the cow's teats, so the dairy world was hungry for an alternative. Farmers were constantly experimenting with the use of platforms, pits, and whatever else they could think of – but nothing caught on.

Ron Sharp (b. 1919, d. 2004) was a dairy farmer near Gordonton, north of Hamilton. He combined the idea of a 'pit parlour', where the farmer stood with the cow's teats above waist height, with the angle parking he'd seen in the main street of Hamilton – and came up with the herringbone shed. If you've ever been to a dairy farm you'll know what I'm talking about because Sharp's invention is now the standard in dairy sheds worldwide.

A long central pit is flanked by two rows of 'bails' – parking spaces for the cows. The cows enter the bails in batches and the farmer walks up and down the pit, attaching the cups to the teats (and keeping an eye out for the tell-tale lift of a cow's tail). When one batch is finished, they are released out the opposite end of the shed and the next batch takes its place. Now one person could milk 75 cows per hour and the maximum practical herd size for a family farm increased from 100 to 400 plus.

There are tales of the odd herringbone-style shed having popped up before this time in Australia, but these were isolated and, unlike Sharp's design, they didn't cause a sea change in the dairy industry. Sharp's idea spread like wildfire. By the mid 1960s, 70 per cent of new parlours were herringbone style – and not just here, but in Australia and the UK and all over the dairying world. By 1972 half of all farms in New Zealand used the new herringbone design and by the early 1980s it was in over 80 per cent of milking sheds, where it stayed until another New Zealand shed design became popular.

The next iteration in milking shed design was the rotary shed. As early as the 1930s an American had designed the 'Rotolactor', which was a massive construction with room for 50 cows on a rotating platform. Its size and expense meant it was only practical for very large herds and it never caught on, until a New Zealander took the rotating platform idea and invented a simple version that literally revolutionised the industry.

An engineer mate agreed to build it for him; and so 3000 of New Zealand's brand new 'dollars' later, on 2 September 1969, Hicks' cows took the new Turn-Style shed for a spin.

In 1967 Eltham farmer Merv Hicks was milking his herd in an old walk-through shed when a dairy inspector gave him two years to upgrade or shut down. Hicks wasn't going to build a herringbone shed because he'd observed that the close proximity of the cows could make them antsy. He began to think of ways to keep the cows separate, and came up with his Turn-Style rotary shed.

An engineer mate agreed to build it for him; and 3000 of New Zealand's brand new 'dollars' later, on 2 September 1969, Hicks' cows took the new Turn-Style shed for a spin.

Hicks' major innovation, which allowed a much simpler shed and cheaper construction, was to have the cows walk forwards onto the rotating platform, then backwards off the platform after a full turn. Turns out the cows didn't mind the backing, and yes, they were immediately calmer than in a herringbone shed.

Hicks set up a company and built thousands of Turn-Style sheds over the next 20 years, then he sold the patents to international dairy giant DeLaval, who sells them internationally.

Today over 40 per cent of New Zealand's dairy herd are milked in rotary sheds – and that's a huge chunk of New Zealand's GDP. More than that, now the whole world knows that the 20th century's two major advances in dairy technology – the rotary and herringbone parlours – come from New Zealand. It's a good look, and it's all thanks to Hicks and Sharp.

If you want to see the original herringbone shed, it still stands near Taupiri and you can make a pilgrimage. If you want to see Merv Hicks, you'll have to track him down on the Tauranga kiwifruit orchard he bought when he sold up the farm. Hicks was given a Lifetime Achievement Award at the Dairy Awards in 2004, while Sharp was honoured with an ONZM in the 2000 Queen's Birthday Honours.

Deer Farming

TALL FENCES AND LOTS OF BAMBI QUIPS MAKE FOR A LUCRATIVE NEW INDUSTRY

People have been hunting deer for thousands of years, chiefly for food and velvet. However, it wasn't until the 1970s that a handful of New Zealanders decided that trudging through the bush for days, aiming at something in the distance which more often than not turns out to be another hunter, then getting lost, getting hypothermia, getting in trouble with Search and Rescue and getting a ride in a rescue helicopter was not an ideal situation. Better to put all the deer in one place and look after them: so deer farming was invented.

Human beings depend on eating to survive and the most efficient way to get meat to eat is to domesticate, breed and then kill some sort of

85

Today, approximately 85 per cent of all venison served in restaurants in the US comes from New Zealand.

Left: Kiwi icon Colin Murdoch (see page 48) with a captured deer. Far left: Domesticated red deer (note the eartags)

animal species. Domestication means individual animals don't have to be hunted, leaving the society more time for other activities. Throughout human history, every breed of animal has been tested for domestication in a process of trial and error. 10,000 years ago sheep, goats and pigs were domesticated, then cows 8000 years ago. Next were horses, donkeys and water buffalo. And by about 5000 years ago, the llama, camel, yak and reindeer were also tamed (along with banteng and the gaur, but I've never heard of those). For 5000 years after that, no further (large) mammals were domesticated. None at all. And it wasn't because of a lack of trying.

Unlike sheep, cattle and other animals, deer do not always get along with each other. They herd during part of the year, but during mating season they are very territorial. They also display behaviour that makes it inconvenient to keep them in one place. While they are not an inherently nasty animal (which is what has kept the zebra safe from domestication), they do show a tendency, like antelopes and gazelles, to panic and scatter when frightened. Discovering the way around these difficulties was what made farming deer possible, and not only possible, but practical, efficient and worthwhile.

In 1973 Dr Ken Drew and Ministry of Agriculture and Fisheries vet Les Porter borrowed a few deer and set up an experimental farm. The two set out to put a structure around deer farming to formulate the basic management practices. They were generally considered 'crackpots', but they persevered, coming up with a set of practices and approaches to deer farming that covered things such as: the right subspecies to choose (in New Zealand we typically have red deer, red deer–wapiti cross and fallow deer); the layout of paddocks; how many deer to keep together; creating a more natural environment; general animal health; and even antler removal.

It's not exaggerating to say that together, Drew and Porter were successful in domesticating the first new species in 5000 years. They also played a large part in establishing deer-farming in New Zealand. While the numbers have fluctuated over the late 2000s, today there are an estimated 1.8 million farmed deer in New Zealand, of which only 600,000 are (very tired-looking) males.

Our practice has spread to other parts of the world, but today New Zealand still has the largest and most advanced deer farming industry in the world.

Recently some enterprising folk in Otago have even decided to create cheese from deer milk. Venison producer John Falkner and cheese maker Simon Berry will have a few hurdles to contend with – deer have very small udders – but they hope this world-leading breakthrough will lead to global success, as deer milk is rich in minerals and omega-3.

Global success is certainly a term we can apply to deer farming in general. Today, approximately 85 per cent of all venison served in restaurants in the US comes from New Zealand. So successful was New Zealand venison, that an 'appellation', or a kind of quality-controlled brand name, has been created. Now we don't eat 'venison', we eat 'cervena'. Whatever the marketing department want to call it, $260+ million a year in exports (plus about $50 million worth of velvet) is a lot of Bambi steaks.

Perendale, Drysdale and Corriedale

HALF SHEEP, HALF…
ANOTHER TYPE OF SHEEP

New Zealanders didn't invent sheep, although we would have if it weren't for the small stumbling block of them having already been invented. Nonetheless, sheep were brought to New Zealand early and eventually thrived – although the first sheep were two luckless merino brought here by Cook in 1773, and dead by 1774. For most of our history, sheep have formed the backbone of the New Zealand economy, the heart of our farming culture and the leg of our Sunday roasts. Say 'New Zealand' to a foreigner and chances are they'll say either 'hobbits' or 'sheep'. In a great example of the change to our economy, wool made up almost 90 per cent of our total exports in 1860; today it's more like 2.5 per cent. Even so, New Zealand remains one of the world's largest exporters of wool and one of the top three producers in the world. Some of our earliest inventions related to sheep, including the ingenious 'sheep dipping bath' invented in the Hawke's Bay in 1908 by Archibald McLean and John Macfarlane, which could take 15,000 sheep a day.

Despite the ready-made varieties of sheep available to us, and in the true spirit of Kiwi ingenuity, we decided to invent our own flavours to suit our own geography. For example, the New Zealand Romney is a distinct breed developed from the English Romney to form the basis of our meat and wool industry. It accounts for over 50 per cent of our national flock.

Geoffrey Sylvester Peren (b. 1892, d. 1980) was a soldier turned sheep researcher at Massey Agricultural College in the 1950s, where he invented

5

While its popularity suggests it is the second most significant breed in the world after merinos, it now makes up only about 5 per cent of New Zealand's national flock.

Above: Sir Geoffrey Peren with the first Perendale sheep. Opposite: Professor Dry, with a Drysdale ram.

a crossbreed of Cheviot and Romney to meet the needs of hill-country farmers on developing land. The Perendale is a hardy, low-maintenance kind of all-terrain sheep for either wool or meat production. Peren's genetic innovation was a resounding success. At one stage there were over 10 million Perendale sheep in New Zealand and in 1975 the breed was officially recognised by Australia. Peren was later knighted for his achievements.

The Drysdale sheep was originated by another Massey staff member, Professor Francis W. Dry (b. 1891, d. 1979), who discovered the sheep gene that made the wool of some Romneys particularly coarse. Formerly these sheep were culled as being too 'hairy' – not fine enough. Professor Dry teamed up with an English company who wanted the wool from these extra-hairy sheep for carpets, and developed the breed. Because he was breeding sheep that might taint the national Romney flock, Dry's programme was rigorously curtailed by the government of the time until demand for the flock grew. Now New Zealand and Australia both have considerable flocks and Drysdale wool carpets are used in computing environments where static electricity is a problem.

But New Zealand's most successful sheep invention is undoubtedly the Corriedale. The Corriedale is a large-framed, hornless sheep, with dark pigmented skin on its nostrils and lips and a heavy fleece of long-stapled, bulky wool. The Corriedale was developed in New Zealand and Australia during the late 1800s by crossing Lincoln or Leicester rams with merino females. A dual-purpose wool and meat breed, the Corriedale is now distributed worldwide. It makes up the greatest population of all sheep in South America and it thrives throughout Asia, North America and South Africa. While its popularity suggests it is the second most significant breed in the world after merinos, it now makes up only about 5 per cent of New Zealand's national flock.

Perhaps most interestingly – particularly if you're not a fan of sheep genealogy – the Corriedale was not invented by Mr Corrie, but by James Little (b. 1834, d. 1885). Little eschewed the opportunity to name his creation the Littledale and instead named it after Corriedale, a property in North Otago which Little had under the supervision of one William Soltau Davidson, New Zealand's refrigerated shipping pioneer – more on him later.

As well as producing our own breeds of sheep, the wool industry has been the source of many other New Zealand innovations, including William Henry Broome's (b. 1873, d. 1943) secret waterproofing formula for his woollen shirts. In 1913 Broome trademarked a name for his rugged workshirts, and today woollen clothes are sold around the world under that name – Swanndri.

And we can't leave a section on sheep and wool without mentioning the merino and the spectacular global success that this sheep breed has enabled – not that merino was a New Zealand invention. Indeed, while the imported breed was initially one of the mainstays of the New Zealand flock, farmers realised quickly that it wasn't good eating, so today it makes up only around 10 per cent of the total New Zealand flock. What it is good for, though, is producing fine wool, which has great properties when made into clothing. New Zealander Jeremy Moon (b. 1970) discovered these properties in the mid 1990s and developed a clothing range and a brand – Icebreaker – which has gone on to be one of New Zealand's largest export companies. Now merino clothing is firmly established as a brand category of high-performance, good-looking clothing, and Icebreaker is one of the world leaders in merino clothing, helping a supply chain of New Zealand farmers turn grass into wool into clothes into dollars for New Zealand. Global success.

'Improvements in and relating to milking apparatus'

THE MOST OVERUSED PHRASE AT THE NEW ZEALAND PATENT OFFICE

5

When it takes, say, five minutes to milk a cow by hand, and you've got 25 of them to milk twice a day, that's close to four hours of back-breaking work.

Right: Melotte was apparently the king of cream separators (1903).

It used to be that if you wanted milk from a cow, you got down there with a bucket and milked it by hand. No problem if you just wanted a quick drink or to feed the family, but as a commercial exercise it was dismal. When it takes, say, five minutes to milk a cow by hand, and you've got 25 of them to milk twice a day, that's close to four hours of back-breaking work, huddled in a damp, cold milking shed. There had to be a better way.

And of course there was. Mechanised milking apparatus first came to the market in Scotland around 1889, though by all accounts these first machines offered little improvement on the old ways. The earliest machine took 20 minutes to milk a cow, but it did milk two at a time, and it saved the milker bending over for the whole process. Further, some of the early machines had cups that would attach to the cow's teats and then not come off again! The teat would swell and the poor cow would be stuck until the farmer loosened a rubber band on the top of the cup. Most unsatisfactory.

In the 1880s in New Zealand, the establishment of refrigerated shipping (see page 258) and the opening of the UK market meant a sudden massive increase in the amount of dairy farming all over the country to meet the demand. The first milking machines from Scotland were in use here, but their deficiencies meant the race was on to develop the best milking machine to extract the white gold from the country's growing dairy herd.

One early example of New Zealand's farmer/engineer breed was John Blake, of Otakeho, Taranaki. He bought one of the earliest milking machines, but quickly realised its limitations. In true Kiwi spirit, and thanks to his training as an engineer, he set about improving the machine, redesigning the cups so the cow could be milked in a quarter of the time.

Word got around his neighbours of what Blake had done, and others asked him to create improved cups for their machines too, which Blake did happily – the extra money supplementing the farm income. However, when he had made only half a dozen sets, the manufacturers of the original machine found out what Blake was doing. Instead of congratulating him on his initiative and purchasing the design from him, they claimed his 'improvement' was an infringement of their patent rights, and forced him not only to purchase back all the cups he'd made for his neighbours, but to cease using the design on his own machine! Rightly annoyed, Blake took the opportunity to design and create an entire milking machine of his own from scratch, a process he finished in 1907 when he sold his first 'Simplex Milker'.

Blake's new design was even better than his last, and the machine was simpler and more efficient than any around. Farmers came from miles around to see the machine in action, and eventually Blake gave up dairy farming to concentrate on selling his milker. Then in 1910 he sold his patent rights, and the machine began to be marketed worldwide.

The first milking machines from Scotland were in use here, but their deficiencies meant the race was on to develop the best milking machine to extract the white gold from the country's growing dairy herd.

Other New Zealand inventors took up the challenge, and creating improvements to milking machines became a national pastime. Sidney Knapp, a dairy farmer from Greytown in the 1920s, was quite an inventor too. Among such inventions as his 'improved hoisting apparatus', 'petrol saving engine' and most notably the 'Knapp sack sprayer' – a backpack flame-thrower unit – were a number of inventions pertaining to milking. Like Blake, he started out by improving elements of the machine, using his steam engineer skills to patent an 'improvement to milking valves' in 1921. And then, like Blake, he went on to create a milking machine from scratch.

In the first half of the twentieth century, more than 36 different brands of New Zealand milking machine came to market, many with features that were unique to New Zealand. These companies, with names like Gane, Ridd, and Eureka, enjoyed varying amounts of success.

New Zealand still sells innovative milking systems around the world. Most of our thousands of milking-based patents are now registered by companies, but there are still one or two individual inventors working away in their milking sheds who keep coming up with 'improvements in and relating to milking apparatus'.

As a final aside, it may interest you to know that the latest European innovations patented are milking robots – a machine that will automatically lead the cow from the yard into the milking stalls, clean the teats, work out from the computer chip in the cow's eartag how best to milk the animal, apply the cups and milk the cow with all the loving care that a computer-controlled collection of metal, wires and electronic components can muster. These are even integrated into systems that allow the cow to decide when it wants to be milked and wander into the milking sheds at its leisure. Ah well, there's no stopping progress.

Animal Eartags

KEEPING TAGS ON YOUR STOCK

Sometimes inventions aren't cool or glamorous. Sometimes they don't seem very ground-breaking and they aren't exciting to teenagers. Unfortunately, the way with inventions is that the ones that are the least fun to look at are often the ones that are the most fun to make money from. Allflex stock tags are one of those inventions.

Put simply, the Allflex tag system is a farm-management tool. It is not an invention that can be described as a first, but it earns its place in this book by being an innovation that made their product easily the best. Before the Allflex system was developed, stock were identified by either a metal tag or a one-piece plastic tag. These tags weren't

1964

In 1964 a Taranaki farmer called Brian Murphy conceived the idea of a new system of cattle identification. He went to John Burford at Delta Plastics to see if he could make it.

Above: A male Allflex tag (blue) hides among the female tags.

easy to see, they rusted, they fell off. The Allflex system solved those problems and enabled farmers to improve the efficiency of their work.

The importance of stock control should not be underestimated. Some historians consider that the need to count animals was the whole reason humans invented numbers. So the first stock control invention was numbers and the latest stock control inventions are being made by Allflex. At first the farmer's needs were simple: how many cows have I got? Now stock control and herd management as well as government-mandated stock identification make it crucial to have information on a pig's ear.

How does stock identification help farmers? Imagine you are being chased full tilt through a field by a bull. The bull is running you down, its hooves heavy on the ground, its red eyes crazed. You rush headlong towards the fence, but can you possibly reach it before the bull gores you? Looking over your shoulder you see by the two-piece flexible polyurethane tags through the bull's left ear that it is a male. You knew that already. The front tag is blue, so it was born in 2011; and if you could see that back tag you would know by the colour that it was one of your own bulls. The bull lowers its heaving head and you feel its hot breath. As your life flashes in front of your eyes, you pull out your scanner and read the microchip and the bull's life story flashes before your eyes as well – you know about its parents, that it was a twin, what percentage pure-bred it is. Just before you are raised on the bull's razor-sharp horns, you give a little laugh. The bull's semen was underperforming and he was due to be culled, but now he's culling you. 'Ah, those ironic bulls!' is your last thought. How much more exciting could an invention be?

In 1964 a Taranaki farmer called Brian Murphy conceived the idea of a new system of cattle identification and went to John Burford at Delta Plastics to see if he could make it. The Duotag and its applicator were the result. Delta and the company that sold the tags eventually merged to become Allflex. The product was immediately better than anything else on the market, but in 1971 the use of the 'male tag' (the Allflex tag comes in two pieces – one with a prong and the other with a hole) in combination with the applicator to pierce the ear and apply the tags in one operation, made the system unique and patentable. It was patented worldwide, and successful exporting commenced.

The product was immediately better than anything else on the market, but in 1971 the use of the 'male tag' ... in combination with the applicator to pierce the ear and apply the tags in one operation, made the system unique and patentable.

Such a simple system is bound to have imitators, and throughout the 1970s Allflex fought legal battles in the United States against patent infringements. Finally in 1982 they won those battles and were awarded punitive damages. Through the 1980s and 1990s Allflex developed electronic eartags to add a new digital-age dimension to livestock control.

In a great example of Kiwi ingenuity becoming a global success story, Allflex products are now manufactured in New Zealand, Brazil, Scotland, Poland, Canada and Turkey, and marketed in all major countries, making Allflex the world's most widely used animal identification eartag. Orders of 10 million plus tags are not uncommon and the tags have been used not only on sheep and cattle but also on fur seals, bears, fish, penguins and even lampposts (which are very disappointing when turned into steaks). Allflex are the self-professed 'Number one livestock identification company in the world' – and why not, when they make something as romantic-sounding as a system for the 'revolutionary electronic tracking of animals from birth to slaughter'?

Why would such a successful agricultural innovation arise in New Zealand? Maybe it is because our distance from the rest of the world gives us a sort of permission to try doing things our own way, to depart from the rules and norms that seem so far away that they simply no longer apply. Or maybe we just like eartags – a lot.

Cross Slot Seed Drill

NO TILLAGE REQUIRED

30

After 30 years of development, Baker began exporting his machines – recognised by many as the most advanced in the world.

Right: The Cross Slot seed drill places seeds on the left and fertiliser on the right of a cross-shaped slot in the earth, and all without major disturbance of the soil.

The prize for the New Zealand inventor with the greatest stamina and passion belongs to agricultural scientist Dr John Baker. In 1967, Baker invented a new technology for sowing seeds and almost half a century later he is still championing it because he believes it represents nothing less than the solution to the world's food needs as Earth's population heads towards 10 billion.

The traditional method of ploughing a field every time you sow is destructive to the land. No-tillage sowing was invented in England: it means you don't lose precious moisture and organic materials and microorganisms when sowing. Working at Massey University, Baker invented a no-tillage sowing machine called the Cross Slot that deposits seed and fertiliser separately into a horizontal slot beneath the surface, then closes the earth back over it. 'Ploughing is like invasive surgery, no-tillage is keyhole,' he says.

After 30 years of development, Baker began exporting his machines – recognised by many as the most advanced in the world – in 1997. It has been frustrating raising money to develop such high-cost drills (between $200,000 and $600,000) and bring them to an international market often suspicious of something new. Baker exports his big, beautiful machines to 17 countries and is hopeful the world will catch up with his innovations. 'Whoever invented the saying "If it sounds too good to be true, it probably is" did us no favours,' he says.

The Aquarium Tunnel

OBJECTS MAY BE BIGGER THAN THEY APPEAR

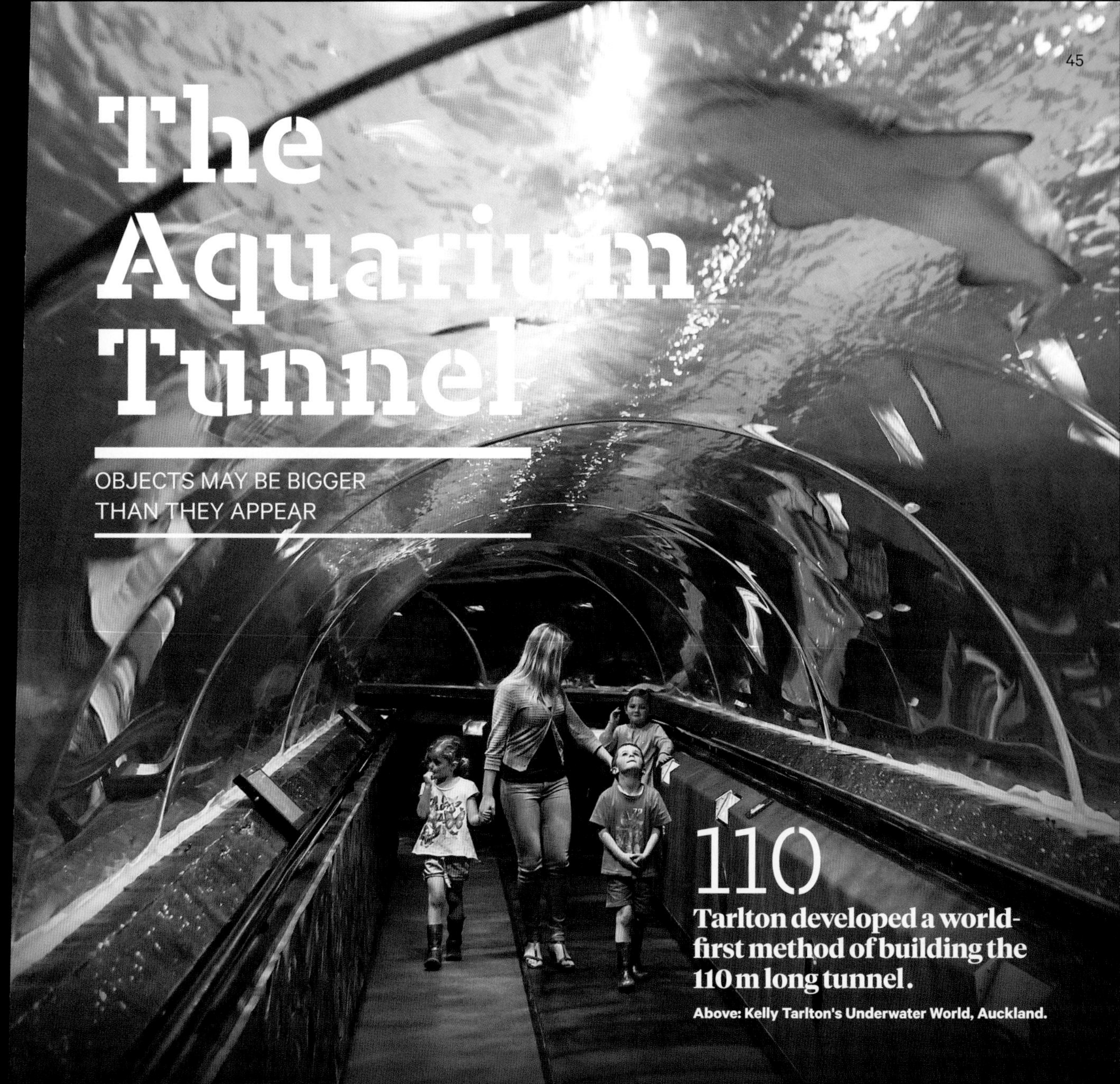

110

Tarlton developed a world-first method of building the 110 m long tunnel.

Above: Kelly Tarlton's Underwater World, Auckland.

Kelly Tarlton (b. 1937, d. 1985) was a scientist, marine archaeologist and diver whose love of the sea was inspired by French marine scientist Jacques Cousteau, inventor of scuba diving equipment – which Tarlton himself later helped to improve. To locals and visitors to Auckland, however, his name is synonymous with the tourist attraction just around the waterfront from the city.

Standing on a conveyor belt at Kelly Tarlton's Sea Life Aquarium, you are transported through an aquarium, staring up at the fish who largely ignore you, through a thick see-through acrylic tunnel. If you read the walls, you'll discover the attraction was built in underground sewage tanks (disused, promise!) and that the tunnels were constructed by Tarlton himself. Indeed, to get the seamless, see-through effect he was after, Tarlton developed a world-first method of building the 110 m long tunnel by taking large sheets of clear acrylic, cutting them to size and heating them in an oven until they took the shape of the mould. Some of the individual sheets weighed over 1 tonne. The tunnel distorts vision so that the fish appear about a third smaller than they really are, but the overall effect is impressive, allowing the public a close-up view of the fish life Tarlton loved. The process and idea have been adopted in aquaria around the world.

The Vacreator

TAKING THE STINK OUT OF BUTTER

Butter smells bad. We just don't know it. And the people we have to thank for our blissful, odourless ignorance are Lamont Murray and Frank Board.

In 1923 Murray and Board were about to open their own butter factory in Te Aroha, and were unhappy about the method that was being used to pasteurise the cream for the butter. Back then cream was often contaminated with outside flavours: when the cow ate strong-flavoured weeds or grass, the flavour would find its way into the milk. There's nothing worse than onion grass-flavoured cream! The established method of pasteurisation (basically boiling the cream and cooling it again) may have killed all the bad things in the cream, but it didn't help it to smell or taste better – it was common for cream to come out of the process tasting cooked or even burnt.

Quite apart from the taste, there were financial and economic reasons to improve the process too. Firstly, the government at the time had regulated it so that suppliers were to be paid on the quality of the cream delivered to the factory, meaning farmers and suppliers had great incentive to improve their product. Secondly, poor quality cream results in poor quality butter, and this was a time when New Zealand butter was beginning to be exported in large quantities to England. The poor quality stuff didn't last as well on the trip, and sold for less when it got there.

Murray and Board were the perfect team to combat the problem – Murray came from a family of milk producers and manufacturers, and Board's family was involved in importing and marketing New Zealand and Australian butter into England. Once they opened their factory in Te Aroha they began immediately to concentrate on how to make their end product better.

1933

By 1933 they had done it. They had come up with a technique for deodorising cream that had no side effects.

Far left: The Vacreator at work, looking shiny and sterile.

By 1933 they had done it. They had come up with a technique for deodorising cream that had no side effects. The technique involved passing hot steam through the cream instead of boiling it. This was quite a novel concept – conventional wisdom at the time frowned on the use of steam, but Murray and Board came up with a way of making it work, and created a series of machines to assist in the process – machines they dubbed the Vacreator.

The Vacreator was a great success. Murray and Board set up a company, Murray Deodorisers, (quite handy – I know a few Murrays who could do with deodorising), to manufacture and license the concept, and it was quickly adopted and used in New Zealand. Board travelled to the US in the 1930s and succeeded in converting the dairy industry there to the techniques the Vacreator offered, diversifying its uses to suit the larger American market.

Today the Vacreator is used worldwide, and it's thanks to Murray and Board that we can enjoy our butter without the need for a nose peg.

The Tranquilliser Gun

IF YOU LOVE SOMETHING, SHOOT IT FULL OF TRANQUILLISER

1959

It wasn't until 1959 that the tranquilliser gun was invented and perfected – and it took a New Zealander to manage it.

Right: Our own Daktari of Timaru, Colin Murdoch. Far right: Murdoch's tranquilliser gun set.

If you were alive during the 1970s, you would have seen the TV show *Daktari*. For those who weren't, it was an action/drama series about a vet on the African savannah, pulling thorns from lions' paws and shooting zebras with a tranquilliser gun. How pleasing then, to learn that the show's main character, Daktari (Doctor) Tracy, wouldn't have been half the swashbuckling man-about-jungle that he was without the help of a New Zealander.

Say you're a vet, a ranger, a zoologist or animal physiologist keen to study animals without killing them. Before the invention of the tranquilliser gun and darts, the only way to catch a live animal without killing it was to chase it down in a vehicle, lasso it with a pole and a loop of rope and wrestle it to the ground where it could be safely injected. If the animal was a dangerous one then even this method would have been too risky, and the animal would need to be trapped. It wasn't until 1959 that the tranquilliser gun was invented and perfected – and it took a New Zealander to manage it.

Let's have a quick look at the difficulties the task poses. Firstly, you have to deliver a dose of tranquilliser into the body of the animal, so you need a syringe

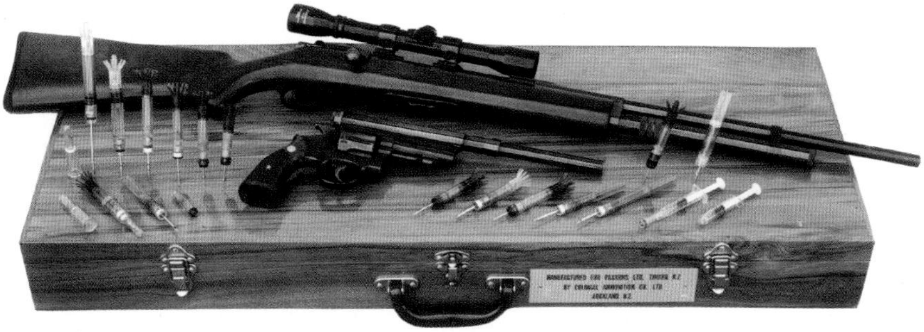

that doubles as a projectile and delivers the dose upon impact. Next, you have to be able to control the velocity of the flying drugs so that, no matter how far away you are from the animal, the needle just sticks into the skin without breaking a bone. This invention was waiting for an expert in gun making, animal physiology, ballistics, pharmaceuticals and anaesthesia.

Colin Murdoch (b. 1929, d. 2008) of Timaru was working with colleagues studying introduced Himalayan tahr (wild goat–antelope!) populations in New Zealand, and had the idea that if a dose of tranquilliser could be safely projected into the animals they would be a lot easier to catch, examine, tag and release.

Murdoch was a pharmacist who had a veterinary practice on the side. During World War II, rifles and shotguns were not imported into New Zealand, so as a serious hunter Colin became an expert at fixing and modifying guns. With the motive and the means, he began to develop the range of tranquilliser rifles, pistols and darts that revolutionised the way animals were studied and treated all over the world.

At first he tested the system on dog-tucker rams in Timaru, then he travelled the world looking for unsuspecting animals to test his invention on. From long range he anaesthetised hundreds of kangaroos, zebras, crocodiles and other animals.

From the moment he produced the first gun in New Zealand, word spread quickly. He patented it and started getting enquiries from overseas. Murdoch's company, Paxarms (pax = peace + arms), began exporting the systems from Timaru to over 150 countries. Every time they sent one to an area, other zoologists got to hear about it and demand grew.

Here's how it works. The guns are modified pistols or rifles, with interchangeable barrels for different calibre syringe projectiles. The power is provided by specially manufactured blank cartridges. Each gun has a dial which can be set to any of 32 positions. The dial controls a valve forward of the pre-expansion chamber and between the cartridge and the projectile.

This invention was waiting for an expert in gun making, animal physiology, ballistics, pharmaceuticals and anaesthesia.

The more the valve is opened, the further the projectile will fly. With the valve at minimum you can shoot an animal in a cage a few feet away; at maximum it will fire 150 metres.

Using a special scope that Murdoch also invented (and which was snaffled by NATO in 1974 for their troops) the range of the animal can be determined and the angle of elevation of firing can be adjusted to control the velocity at which the syringe hits the animal; fast for thick hides, slow for thin hides.

The syringes themselves are made of incredibly strong, light polycarbonate plastic with a needle threaded onto the front, and a small shuttlecock-like tail on the back to give very accurate shooting. On impact a special needle and valve system lets a free piston inside the syringe push the dose into the animal. It's all brilliant.

Indeed, the gun built upon Murdoch's previous success; he had invented and patented the plastic disposable hypodermic syringe (see page 55). In all, Murdoch had 46 patents for his inventions, including a silent burglar alarm and another type of childproof lid (see also Claudio Petronelli, page 242). That said, he commented himself that while patents gave you the *right* to sue someone for copying your idea, they didn't give you the *money* to sue. In 2000, Murdoch was appointed an Officer of the New Zealand Order of Merit. He eventually sold Paxarms to lighten his workload, then retired and built himself a yacht.

So next time you're crouched over a heavily sedated cheetah, administering much-needed dental work and then watching him awake, none the worse for wear, to wander off majestically into a safari sunset, think of Colin Murdoch – the Timaru Daktari who made it all possible.

3.

The Human Body

DNA, VALVE REPLACEMENTS
AND A BIONIC MAN

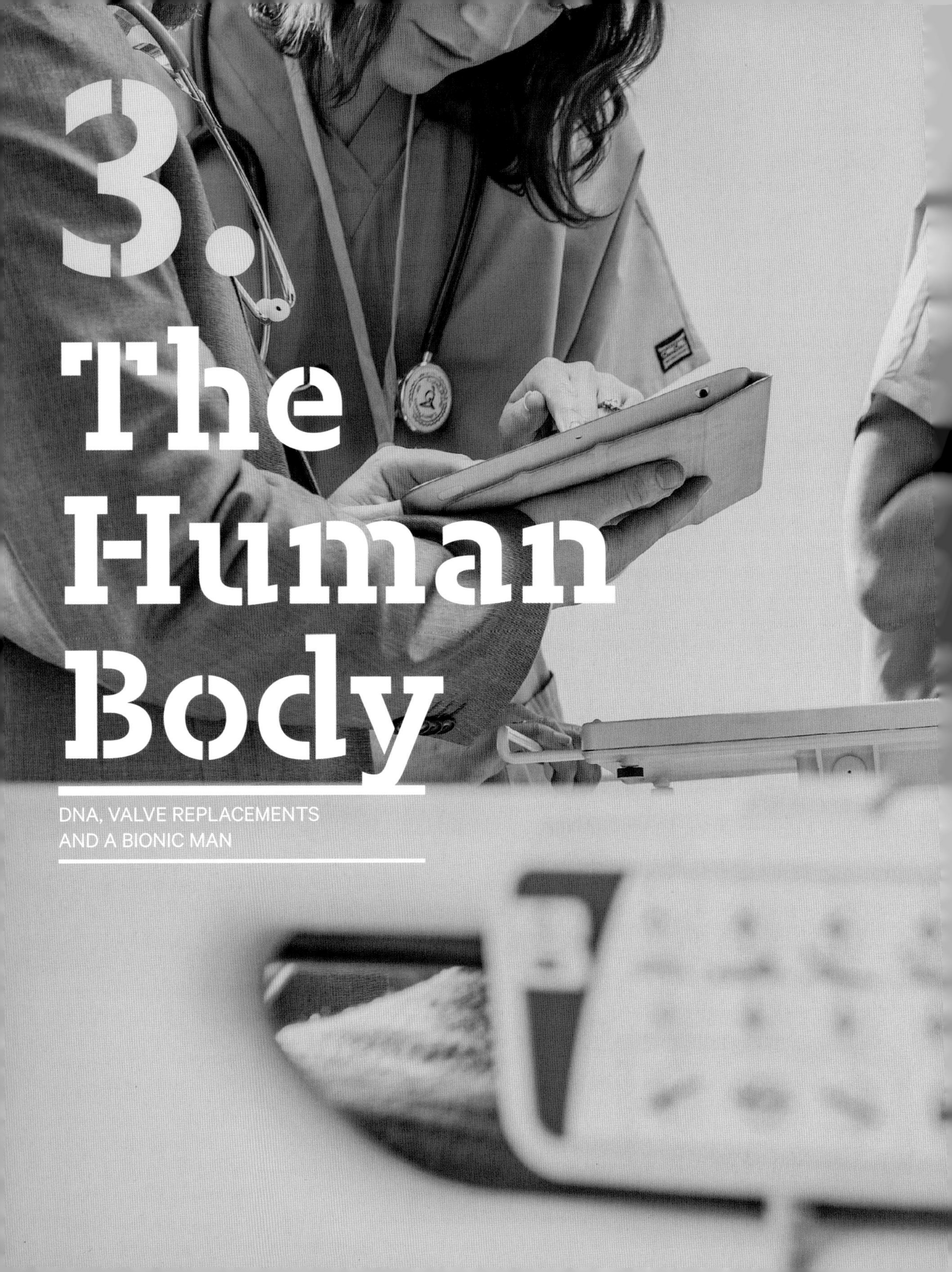

There were far too many brilliant Kiwi inventions to fit into this chapter. One inventor at age 88 came up with a way for 'poor people' (as he put it) to make their own bidet. In his estimation, not being a millionaire was no reason to have a dirty bottom.

An innovative Auckland company came up with a device for helping stretch the lower back, surely something that warrants a 15-minute advertisement on TV in the mornings if ever there was one. And Tapanui flu could be said to be a New Zealand invention – this illness affects people not just in the West Otago region, but all over the world.

It's easy to see why New Zealand inventors try to make inventions for the human body. Not everyone in the world has farms, lakes, beaches or roads just like ours, but wherever you go, *everybody* has a body.

Bionic Man

COSTING EVEN LESS THAN
SIX MILLION DOLLARS

2003

In 2003, when Scottish-born Kiwi import Robert Irving was diagnosed with multiple sclerosis, he knew that a future in a wheelchair might be ahead of him. As an engineer, however, he decided to do something to prepare for this eventuality.

Right: The bionic legs in action.
Far right: Richard Little and Robert Irving of Rex Bionics.

The idea of making a bionic man (or woman) seems like a fanciful one. If you were around in the 1980s you will have seen Steve Austin, 'a man barely alive', have his body reconstituted using a mix of robotics and electronics, in the TV show *The Six Million Dollar Man*. A generation of kids watched the show and the antics of the hero, and probably thought the whole thing too far-fetched to ever really happen.

While the idea of having X-ray eyesight and running 100km/h might still be the stuff of science fiction, the fact is that bionics for humans are no longer just a dream – indeed, thanks to the team at Rex Bionics of Auckland, the Six Million Dollar Man is now heavily reduced to just $150,000. 'Bionics' is the replacement or enhancement of body parts by mechanical versions, often surpassing the original function – mixing machines and (in Rex's case) humans to design enhancements. Sounds spooky, but if you are a wheelchair-bound accident victim you will be interested to hear that Rex's 'exoskeleton' can mean an opportunity to stand and walk again – even to climb stairs.

Little says he can't imagine that in 10 or 20 years' time we will still be giving people wheelchairs.

In 2003, Scottish-born Kiwi import Robert Irving was diagnosed with multiple sclerosis. His own mother was already a sufferer of the debilitating disease so he knew that a future in a wheelchair might be ahead of him. As an engineer, however, he decided to do something to prepare for this eventuality. He teamed up with fellow engineer and another import to New Zealand, Richard Little – the two describe themselves as 'Kiwi by choice' – and they decided to use their engineering know-how to design and develop a machine that would allow wheelchair users to stand and walk.

Their first design sketch was decidedly low tech, literally on the back of a beer mat, but the machinery they put together was anything but. They describe New Zealand as a great place to innovate – an easy place to do business, with lots of innovation nearby: 'We googled it, and discovered the world leader for a component we needed was in Onehunga.'

The robot they developed is a complex mixture of computer hardware, software, electronics and electro-mechanical parts, expertly assembled and designed to fit around the legs and lower torso of the body. From a sitting position, where users strap themselves in, the bionic legs can help them stand and walk – and much more besides. The sophisticated machinery allows the user to control it with a small joystick. There are even plans to extend the 'sci-fi' factor further, with a collaboration that would allow users to control the legs with their thoughts alone – harnessing the electrical impulses of the brain and translating them into commands for the Rex machine.

It takes users a little while to get used to them, and they are not currently as fast or agile as flesh legs, but the bionic legs clearly show a future where wheelchair-bound people have an alternative. They are relatively light and very stable – users can stand upright in busy areas, and a small knock will not topple them over. The battery power can last a couple of hours of continuous use, with a short recharge time.

The legs allow a new-found freedom for their users, allowing accident victims, chronic disease sufferers and others to experience the world at the right height again (actually, users are normally a bit taller because of the extra thickness of the footpads). As well as the freedom and self-esteem the legs bring, there are health benefits – the human body is not designed to be sitting constantly, and standing or walking increases blood flow and improves digestion etc. (By 'etc.', we are referring to bowel and bladder movements, but let's keep this above the belt.)

The company is marketing and selling the products around the world. Like all start-up businesses, it's a hard road, even with such an amazing device. What got them through the early years was 'passion and bloody-minded stupidity'. Access to capital was hard, which is why, in early 2014, Rex undertook a 'reverse takeover' worth £9.5 million with a UK company. This is a quick way for a company to become publicly listed, and is an exciting move for the company, which has spent years in R&D mode, and is, as Little puts it, 'there now'.

With thousands of enquiries, Rex Bionics are on the cusp of true international success. They have plenty of ideas on how to further improve their product – already the most sophisticated robot in the world – and see that in 20 years' time, their technology will be obvious. After all, Little says, he can't imagine that in 10 or 20 years' time we will still be giving people wheelchairs. People expect it, and Rex is delivering it: world-leading technology, from two adopted Kiwis at the bottom of the world.

The Cortex Cast

3D PRINTING MADE USEFUL

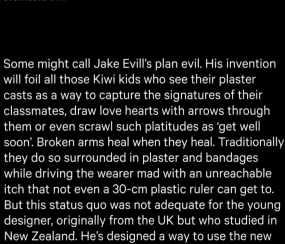

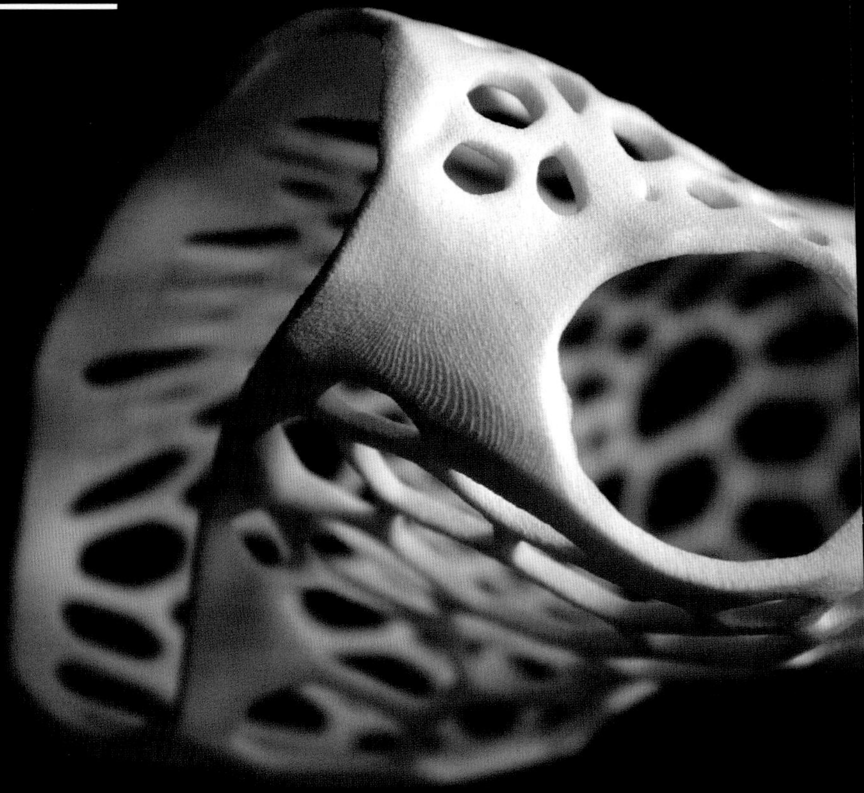

The specific design of each cast is calculated by a special software package, and it creates a unique pattern designed for best fit and optimal support of the key area that needs it.

Right: The Cortex Cast even *looks* evil.

Some might call Jake Evill's plan evil. His invention will foil all those Kiwi kids who see their plaster casts as a way to capture the signatures of their classmates, draw love hearts with arrows through them or even scrawl such platitudes as 'get well soon'. Broken arms heal when they heal. Traditionally they do so surrounded in plaster and bandages while driving the wearer mad with an unreachable itch that not even a 30-cm plastic ruler can get to. But this status quo was not adequate for the young designer, originally from the UK but who studied in New Zealand. He's designed a way to use the new technology of 3D printing, coupled with scanning and X-raying of the broken appendage, to make a perfectly moulded, individually created, lightweight and stylish alternative to the plaster cast. The cast consists of a web-like hard plastic shell, made from nylon, printed in a 3D printer to an individual design. The specific design of each cast is calculated by a special software package, and it creates a unique pattern designed for best fit and optimal support of the key area that needs it. The design won Evill the 2013 James Dyson award for design and innovation, and he is currently looking into ways to commercialise the product and bring it to a waiting world. Hopefully this helps Evill get his big break.

Disposable Syringe

SAVING LIVES SINCE 1956

1961

The 'Disposables Revolution', as it is called, hit the medical world in 1961 and every major manufacturer of syringes in the world has been making their versions of Murdoch's designs since then – billions and billions of them.

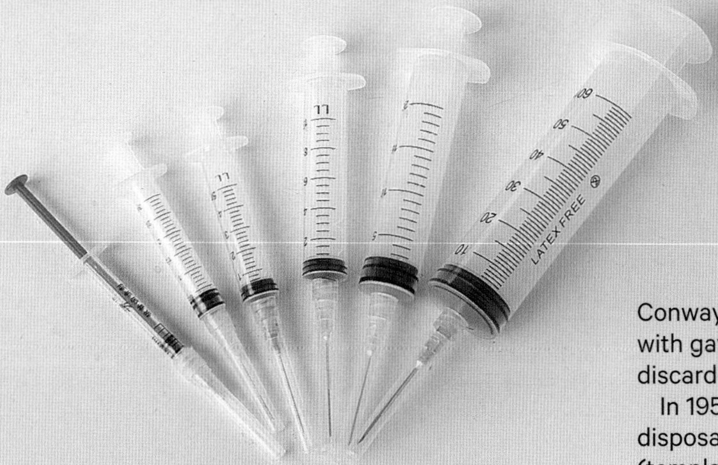

Left: Modern versions of Murdoch's syringe.

If you measure greatness by the effect a person has on the world, Colin Murdoch (see page 48) must be counted as one of our greatest New Zealanders. His invention of the disposable plastic syringe has saved millions of lives and transformed medicine worldwide.

Growing up in Christchurch, his childhood dyslexia made school tricky, but didn't stop him wanting to shoot things. As a 10-year-old he invented his own sort of muzzle-loading pistol using 'acid ignition' and went about the countryside shooting rabbits. At 13 he was driving his own car around North Canterbury and nobody seemed to care; and he became a hero when he saved a man from drowning in the New Brighton estuary. It turned out to be just the first life he would save.

The glass syringes that were used at that time were just sterilised between uses, but often the diseases spread by improperly sterilised syringes were worse than the diseases they were meant to be curing. Murdoch's idea was simple, and it changed everything.

Flying on a DC3 one day in 1952, Murdoch had a 'eureka moment'. The way the cap fitted on the Conway Stewart fountain pen he always travelled with gave him the idea for a syringe that could be discarded after a single use.

In 1956, at the age of 27, Murdoch patented the disposable syringe. He set about making dies (templates from which syringes could be manufactured) and took a prototype to the New Zealand Health Department to get their blessing and help with its development. They said his idea was good, but 'too futuristic'. In their opinion, nobody would want to be injected with a plastic syringe. Says Murdoch's wife Marilyn: 'Colin came home from that trip pretty dejected.' Murdoch couldn't afford to bring his ideas to market himself but that didn't stop the idea spreading. As he continued to develop variations of his syringes, companies around the world began to make disposable syringes – without paying a cent in royalties to Murdoch. The invention most likely spread by the very patents that Murdoch had published to protect it. The 'Disposables Revolution', as it is called, hit the medical world in 1961 and every major manufacturer of syringes in the world has been making their versions of Murdoch's designs since then – billions and billions of them.

The vaccine is perhaps the greatest advance in medical history, saving hundreds of millions of lives since its introduction. We should celebrate Colin Murdoch, the New Zealander whose syringe made vaccination safe for the world's billions of inhabitants.

Plastic Surgery

THE 'FATHER' OF PLASTIC SURGERY – AND HIS COUSIN

1947

McIndoe was knighted in 1947 'for his remarkable work on restoring the minds and bodies of the burnt young pilots of World War II'.

Right: Sir Archibald McIndoe celebrates the wedding day of a former patient. Far right: Sir Harold Delf Gillies.

Imagine the scene: you are crouched in a trench during World War I. The enemy is just metres away, machine guns and bombs at the ready. The mud squelches around you. You smell the sweat and fear as you hear the order to go over the top. Up you jump, hoping against hope that something keeps you safe from all that flying metal.

For over 5000 Allied soldiers in the fields of France, Belgium and other battlegrounds of the war, luck was not on their side. Facial injuries – usually gunshot wounds – were all too common, and it fell to a young New Zealand-born surgeon to come up with a way to cope with the often horrific results. Harold Delf Gillies (b.1882, d.1960) was born in Dunedin and spent his youth in New Zealand. Son of an MP, Gillies was quite a sportsman, winning prizes for rowing and, in later life, representing England at golf.

Gillies moved to the UK to study at the University of Cambridge. After graduating, he studied surgery in a London hospital and became known as a talented ear, nose and throat surgeon. When war broke out, he joined the Red Cross and saw many injured soldiers who needed facial surgery. He began to think about better ways to help these guys, some of whom were horribly disfigured. Ironically, he cited as one of his earliest influences a German book he was given which outlined the current German thinking in this area of medicine.

Gillies proposed to the British Army that they create a unit for this 'plastic surgery' work, and in 1916 was ordered by the War Office to head up this new unit as a surgeon in the Royal Army Medical Corps. It was here that he saw the opportunity for a number of different ways to treat facial injuries. Working with dentists, other surgeons and a tragically unceasing stream of patients, he pioneered techniques for facial reconstruction.

While such surgery had been around a long time, it was Gillies who pioneered the idea that not only should the reconstruction be functional, it should also be aesthetic. Plastic surgery was an opportunity

to make the person look at least as good as, or even better than, they had before. His colleagues would see photographs of patients taken before they were disfigured, before surgery, and after surgery, and would often exclaim that Gillies had made them look better than before they were injured. Gillies was the first surgeon who drew pictures and models of his patients before and after facial reconstruction, and his pastel drawings have been praised for their artistry.

Gillies pioneered a technique called a 'pedicle tube'. This involved taking sections of skin from patients' chests, rolling them up and stitching them together, while keeping the end of the tube connected to the chest. The other end was then attached to the face, closing off the facial wound, keeping the blood supply going, and allowing the skin to graft together. In this way he could reconstruct noses, eyelids and other parts of the face which had been damaged.

After the war, Gillies published his techniques in *Plastic Surgery of the Face*, which became the seminal work on the new discipline. He established a private practice in London and was knighted in 1930. But Sir Harold's career was far from over. While he continued to heal accident and war victims, a new set of clientele also appeared – people requesting cosmetic surgery. He always considered it secondary to his reconstructive work, but Gillies pioneered techniques in facelifts and became sought after by society figures and film stars. He didn't always agree with the reasons for the surgery, and commented about one customer, 'When she began to look like a Chinese dance-hall hostess I decided that it was time to stop. She stalked off to spoil some other surgeon's reputation.'

Gillies also helped train another pioneer and eventually world-renowned plastic surgeon – his own cousin and fellow Dunedinite, Archibald Hector McIndoe (b. 1900, d. 1960). Their meeting in London was fortuitous – McIndoe having given up a promising career in the US to chase a nonexistent job in London. They eventually formed a partnership in Harley Street and began one of the greatest medical partnerships of the twentieth century.

Together, Gillies and McIndoe created breast reduction and enhancement techniques without which a number of modern Hollywood starlets would not be so noteworthy. While at the time breast surgery was considered frivolous, Gillies and McIndoe considered it not only physically enhancing, but also psychologically positive for their patients' self-esteem. As final proof of his influence Gillies is widely acknowledged as having performed the first surgical sex-change operation (a male to female conversion), and his cousin rather interestingly gives his name to the 'McIndoe Vagina'.

Assisted into the profession by his cousin, McIndoe went on to have his own brilliant career: he discovered that the liver has two sources of blood flow; he became sought after as a fast and skilful surgeon; and in 1945, at the outset of World War II, he followed in his cousin's footsteps and became the chief plastic surgeon for the war effort. In this role, he not only helped a large number of disfigured war casualties, but also pioneered the saline bath as a technique for helping burn victims. As a non-commissioned member of the forces, he was free to challenge the accepted practices of the time and was successful in finding a number of much more humane and effective ways to help victims, both physically and psychologically. McIndoe's base was the burns unit at Queen Victoria Hospital in Sussex. His patients there became known as the 'Guinea Pig Club', because the surgery they underwent – some of it extending over years – was still in its infancy. McIndoe was knighted in 1947 'for his remarkable work on restoring the minds and bodies of the burnt young pilots of World War II'.

Gillies Ave in Auckland was not named for Sir Harold, but for a much earlier judge and police superintendent. However, in a nice piece of symmetry, it is today home to a number of medical practices including those of plastic surgeons who owe their craft to the Dunedin cousins rightly remembered for their individual impact on the lives of so many victims of the wars, and for the techniques and procedures they pioneered.

Needleless Injections

NOT TO BE CONFUSED WITH NEEDLESS INJECTIONS

We don't usually condone one New Zealand inventor putting another out of business but in this case the invention is so good we'll let it go. Colin Murdoch was the brilliant Kiwi who gave the world the plastic disposable syringe (see page 55). Fifty years later Kiwi professor Ian Hunter at Massachusetts Institute of Technology (MIT) and Dr Andrew Taberner at the University of Auckland have ganged up on Murdoch and are about to make his invention obsolete.

Hunter is one of our academic giants. Working most of his career in North American universities (mainly a little place called MIT) he has made hundreds of advances in the studies of instrumentation, biomimetics, microrobotics and medical instruments. He's a serial inventor with over 100 patents and 20 companies to his name. Unfairly, we feature just one of them in this book – but it's a goodie.

20

He's a serial inventor with over 100 patents and 20 companies to his name. Unfairly we feature just one of them in this book – but it's a goodie.

There are three main problems with needles. One of them is 'needlestick injuries' where doctors and nurses accidentally stick themselves – often leading to sickness and even death. Problem two is that syringes can't deliver precise doses. And the third problem is just that needles hurt. The fear of needles causes people – in some cases even fully grown authors – to avoid treatment.

Plenty of people have tried to solve these problems before, but until now inventions to deliver drugs without a needle have been noisy, often painful machines involving springs, compressed gases or explosive chemicals that don't allow sufficient precision.

Hunter's device shoots a jet of liquid into the skin. A computer-controlled electric motor powers the injection and a computer interface precisely controls how much of the drug is shot in and at what speed, thus controlling how deep it goes. Top speed is about the speed of sound. The diameter of the jet of drugs is the same as that of a mosquito's proboscis, and the injection will be as painless as a mosquito bite – without the itchy lump.

It can also deliver a drug directly into the retina and the inner ear – places needles can't go. And it can shoot powdered and very viscous drugs into you, something no syringe can do. This is important because drugs in powdered form are more stable and don't have to be refrigerated. This has the potential to make drugs much easier to distribute, especially to the developing world.

Hunter's MIT lab and the University of Auckland lab are working to develop the technology, partnering with New Zealand and US companies to bring the jet injectors to market and fulfil his vision that needles will be needless (or at least we'll need less).

Stretch Sensors

BUILDING BLOCKS OF HUGE POTENTIAL

2013

In 2013 they set up the StretchSense office where they make the world's most advanced stretchy sensors for the measurement of human motion.

In 2006 Ben O'Brien and Todd Gisby were two young scientists just beginning their research work in Iain Anderson's biomimetics lab at the University of Auckland. For seven years they investigated artificial muscles; but when they started to commercialise their ideas, customers kept asking if they could make sensors. Says O'Brien, 'By the twentieth or fiftieth time we were asked, we suddenly had the realisation, "People want sensors!"' In 2013 they set up the StretchSense office in one big room with a roller door that opens to overlook a rubbish dump, and here they make the world's most advanced stretchy sensors for the measurement of human motion.

StretchSense sensors are small black elastic strips (actually polymeric stretchy capacitors) with electrodes at each end. They are attached to a small battery and an electronics board that transmits the amount of stretch in the band by Bluetooth to a smartphone app where it can be read. Stick one to the back of your finger and when you bend it, the sensor stretches and reports the amount of stretch to the app. StretchSense didn't invent the flexible capacitor technology, nor the battery, nor the smartphone – but in putting them together and developing the patented algorithms to extract the data from the sensors, they've created a world's first simple, comfortable soft sensor. It could have a million uses.

Filmmakers can use them to capture and animate precise 3D movements of hands and other body parts. Hands are difficult with current camera-and-dot technology because the fingers obscure each other. In human–computer interaction, gestures captured with the sensors sewn into a glove, for example, could be used to control a computer in a more complex and intuitive way than a mouse can now. And athletes and doctors working in rehab can use the sensors to precisely monitor the movement of joints.

O'Brien is quick to point out that StretchSense is just providing the building block for companies developing future products. Just as the stretchy technology, the battery, the smartphone and the app store were the building blocks in their own invention. Orders are coming in, and the potential is huge. 'We're not in this to be small. We're in this to be mega-giants. There are many forces pressing towards this mobile, comfortable, ubiquitous digitisation of the self and we want to dominate our part of that: the stretch sensing part.'

The Discovery of DNA

THE BLUEPRINT OF LIFE HAILS FROM PONGAROA

4

In human DNA there are only four different amino acids: adenine, cytosine, guanine and thymine, commonly known by single letters, A, C, G and T.

Right: Wilkins with his model of DNA.

Wilkins' work became the greatest evidence in support of the helical structure of DNA, and further, on its internal make-up.

When the King of Sweden presented the 1962 Nobel Prize for Medicine, he didn't just give it to those limelight-hogging scientists, J. D. Watson and Francis Crick. There was a third recipient, who seems to be often forgotten – New Zealander Maurice Wilkins (b. 1916, d. 2004).

Watson and Crick's contribution to the understanding of deoxyribonucleic acid is well heralded, and if you asked any reasonable educated person who discovered DNA, theirs would be the names put forward. There is a good reason for this: for a long time the idea of the double helix structure of DNA, and its role in life, was referred to as the 'Watson–Crick' conjecture, after the two scientists who basically guessed it first. But it wasn't until Maurice Wilkins had done an awful lot of work that the proposal was accepted as fact. In short, Wilkins 'proved' the conjecture.

Maurice Wilkins was born in Pongaroa in 1916, but moved to England fairly early on in life to be educated. There he attained his PhD in 1940 for work involving phosphorescence, and then went to work during the war on projects such as the improvement of radar screens. He worked for a while separating uranium isotopes, and did a stint on the infamous 'Manhattan Project' in California, until the return to peace in 1945 allowed him to return to his studies.

He worked for many years after the war on the study of various chemicals and structures in cells, honing his techniques and methods until in the late 1950s he began X-ray diffraction studies of DNA and sperm heads. This technique involves firing a whole lot of X-rays at a DNA molecule, and then studying the pattern of 'diffraction', i.e. where the X-rays bounce off. DNA strands were too small to be seen with the most powerful microscopes available at that time, so X-ray diffraction was the only way to try to derive their structure. Wilkins' work became the greatest evidence in support of the helical structure of DNA, and further, on its internal make-up.

Wilkins and another colleague, Rosalind Franklin, showed that DNA is made up of two tiny fibres, coiled around each other in a spiral shape – a helix. If you could see them close up they would look like a spiral staircase. The links between the two strands – the 'stairs' in this analogy – are molecules of chemicals called amino acids.

In human DNA there are only four different amino acids: adenine, cytosine, guanine and thymine, commonly known by single letters, A, C, G and T. Surprisingly, the amino acids can only link to each other in one way, with G & C and T & A always paired.

It is something as simple as the order in which these four amino acids occur which leads to the huge variety of life around us. Where in your DNA structure, in the part that codes for, say, eye colour, you may have the amino acid pattern GCCGCTAGCCCG which might lead to brown eyes, I'll have GAATCCGGATAT, which makes my eyes blue. Of course, these aren't the real patterns, they're probably a lot more complex (the DNA 'formula' for a human being takes about 3×10^9 of these patterns), but in a simple sense this is what Watson, Crick, Franklin and Wilkins showed. It was the work that showed the world how an individual is defined, and for which Wilkins won the Nobel Prize.

Cxbladder

ACCURATELY DETECTING CANCER WITHOUT THE PAINFUL NEEDLE

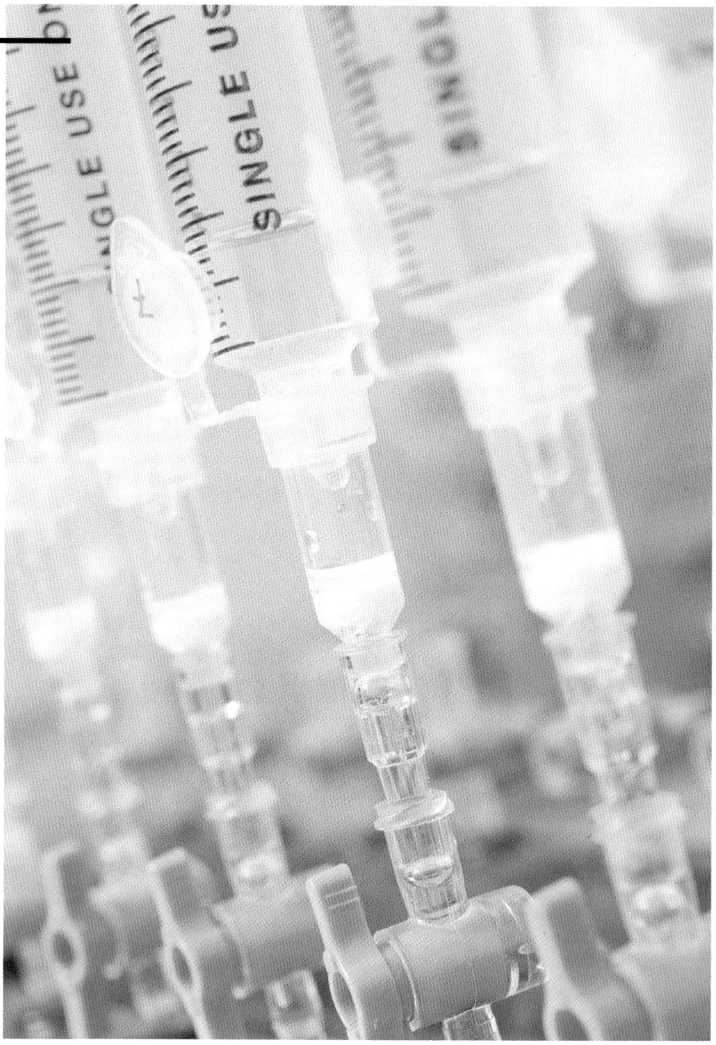

20

It's an amazing process, unthinkable 20 years ago, and a technique that can be adapted to a number of cancers and genetic illnesses.

Far right: Technicians testing samples for cancer markers.

Bladder cancer is the ninth most commonly diagnosed cancer globally. It is much more common in men than women. The main symptom of bladder cancer is blood in the urine, but that doesn't always mean that the patient has cancer – neither does the absence of blood mean it's guaranteed you *don't* have cancer. It's a tricky one to diagnose, but the good news is that if it is detected early, treatment is often successful. The problem is that the prevailing technique for accurately detecting the cancer is painful and embarrassing. And painful again. It involves putting a long thin needle up into your bladder via the... well, the place where your wees comes out. If you just winced, or moved around uncomfortably, then you understand – yes, in *there*.

Pacific Edge Biotechnology Ltd is a Dunedin-based company, formed in 2001 to commercialise research in the area of gene expression profiling, looking into ways to detect abnormalities in genetic samples. What that means is they've worked out a way of

They are seeing test results with 69–90 per cent reliability, and a much lower risk of 'false positives'.

looking at the genetic profile of the patient's cells in order to predict the likelihood of cancer.

Drawing on a tight relationship with the University of Otago, Chief Scientific Officer Professor Parry Guilford realised that their gene expression techniques could be used to detect cancers in the stomach, bowel and bladder. The techniques they pioneered involve detecting certain genetic patterns in a small sample of DNA – in the case of Cxbladder this is picked up from the patient's urine. No need to go in and get it; they can get some when it comes out of its own accord, immediately making the whole thing more palatable (not that that's a word you'd use to describe the collection process).

The basic idea is that cancer in the bladder will form a tumour, and some cells from that tumour will break off and pass out with the urine. The urine is then mixed with a fluid that Pacific Edge created, which extracts the 'messenger RNA' (mRNA) – a family of molecules that carry around information within a cell. Within the mRNA scientists look for signals of five different 'biomarkers' – specific sets of coding which show that this cell can or will do a certain task. Through a series of chemical and biological processes, they can tell if the right combination of these markers is present, which would indicate the cell is likely to be cancerous. It's an amazing process, unthinkable 20 years ago, and a technique that can be adapted to a number of cancers and genetic illnesses. They are seeing test results with 69–90 per cent reliability, and a much lower risk of 'false positives' (detections when there is no cancer). This means a large number of patients can get the test done painlessly and non-invasively, and only those with a heightened risk or a positive result need experience the needle.

While this makes it all sound easy, there was no guarantee of success. They had a good idea, but it was not certain to work, and the team at Pacific Edge laboured for three years to create the database of gene expressions and the first prototype product. Once they saw the first results, they needed three or four more years to commercialise it – getting the right regulatory and compliance approvals. Through this important period of company growth they had support from the New Zealand government and a loyal set of shareholders. The initial phases of such an IP-driven company are very expensive, with large costs in people, IP protection and administration – particularly in compliance.

It's all paying off though. After announcing their product Cxbladder, Pacific Edge became a stock market darling – no surprise really given the CEO is one Dr David Darling. Investors piling in seem to be expecting big returns from the still fledgling company, even though it's earning minimal revenues at the time of writing. In 2013, the stock price tripled in value within a few weeks after a number of years of very flat growth. The company has received a lot of attention, and was the supreme winner at the 2013 New Zealand Innovators Awards. The initial work they put in to understand how genes express themselves is also paying off – they can now diversify into other product areas quickly and keep ahead of the competition.

The sky's the limit for a company like Pacific Edge. Darling says he doesn't spend a lot of time thinking of the exit plan, but instead concentrates on good science, and running a profitable business. Patients the world over are benefiting from that focus.

Smart-inhaler

MEDICINE THAT CAN TELL ON YOU

Advances in medicine have done more to enhance human life than even our amazing advances in reality television. Doctors are better trained and better equipped and pharmaceuticals are more effective. The part of the equation letting things down is the patients. If people don't take their breakthrough medicines when and how they're supposed to, they won't get better.

In 2001 Garth Sutherland invented the Smart-inhaler to help patients lift their game. Smartinhaler attaches to your normal puffer and reminds you when to take a dose, then automatically stores data on when the medicine is taken. This data is wirelessly available to both patient and doctor, who can follow progress and make decisions based on how much medicine the patient actually took rather than how much they were supposed to take.

After 18 months' development Sutherland's company Nexus6 sold its first product to New Zealand scientists doing clinical trials with inhalers, and in 2005 full distribution began around the country. In 2013, with 20,000 Smartinhalers in use around the world, Sutherland raised over $4m to take the technology 'to scale'. Having improved the patients, it's probably now time for Sutherland to do something about the other major problem in medicine – the age of the waiting room magazines.

Mesyn-thes

THIS TREATMENT REQUIRES GUTS

If you don't like the idea of putting squished-up sheep stomach onto you, then you are missing out on a radical new treatment from Auckland company Mesynthes. Ovine forestomach is treated to remove any elements that would make the human body reject it, and then turned into a 'matrix' treatment – like a really flash plaster – that has been proven to have significant benefits in wound treatment. It's so good that the US Food and Drug Administration (FDA) has approved it for human use, opening up an important market for the innovative New Zealand company. Their product Endoform supports the regrowth of human tissue, and can be applied to wounds not only externally but also internally, on ulcers for example. With significant research and IP protection behind them, they are now set to turn this somewhat surprising ingredient into a globally successful product.

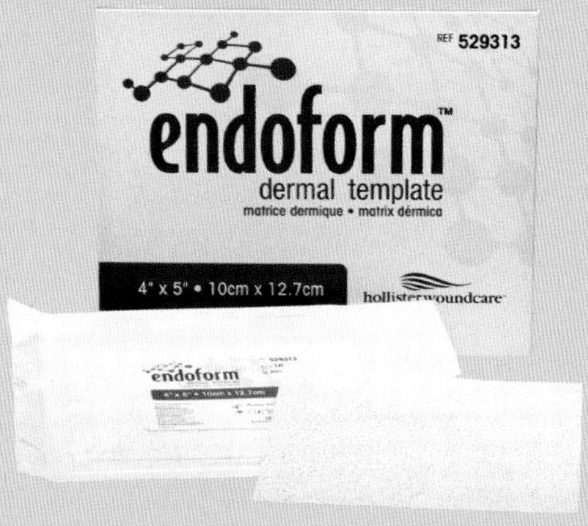

Baby Cooling Cap

PREVENTING BRAIN DAMAGE WITH A CHEAP CAP AND GOOD SCIENCE

Right: The Cooling Cap at work as part of a worldwide trial.

Babies can't wear a helmet when they are being born – but perhaps they'd like one, even if their mothers wouldn't. A baby is more vulnerable during birth than at any other time of their life; there are plenty of things that can go wrong. Lack of oxygen is one big danger. If for any reason oxygen is cut off to the baby they may suffer brain damage, leading to cerebral palsy, disability or death.

One of the issues with reduced oxygen supply is that it's very hard to know until the last minutes of a birth if something is going wrong. The umbilical cord might wrap around the baby's neck, or the womb might rupture. Once the baby is born, oxygen supply is restored, but by that stage the brain may already have sustained damage. About 100 babies a year in New Zealand suffer the long-term consequences of this damage (hypoxic-ischemic encephalopathy or HIE for short). In some countries HIE is a leading reason for birth-related lawsuits, and worldwide, caring for the victims comes with a long-term societal cost.

A Kiwi doctor and scientist, Sir Peter Gluckman and a team of physicians in New Zealand and around the world have invented a treatment for oxygen deprivation in newborns – a small cap that costs just $100. The cap works by pumping cool water around the affected baby's head for the first few days of their life. This cools the head and brain down from the normal 37.8°C to 34°C – the rest of the body is kept warm, but the cooling of the brain puts it into an almost hibernation state. This slows the brain processes and stops the damaged cells from dying off.

A co-ordinated worldwide trial during the mid-2000s showed that the cap significantly reduced long-term damage caused by oxygen deprivation at birth. The viability of the treatment, and the cap, was proven. Gluckman's team's priority was to get this life-saving device out to the world as quickly and cheaply as possible, so they sold the licence to a US company which sells the 'Cool-Caps' at a very accessible price. With the all-important FDA approval now in place, Gluckman's team's invention is set to save thousands of babies and parents a year from a life of hardship.

As with many inventions, the team were initially studying the effects of cooling as part of their research on a completely different topic when they realised that the results they were seeing could be used to prevent brain damage. Since then, other researchers have taken the idea of cooling the body even further, and the technique is now gaining even broader usage in medicine. Gluckman himself went on to further promote the use of science and medicine to the wider public when he became Chief Science Advisor to the Prime Minister (and when he wrote the foreword to this book!).

The First Aortic Valve Replacement

MEDICAL INNOVATION BY A KNIGHT IN STERILE ARMOUR

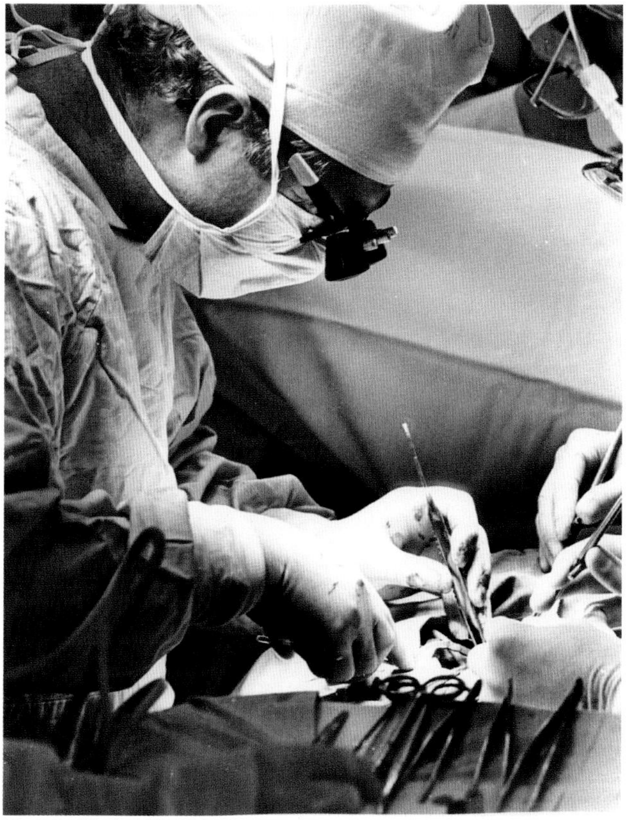

1958

He returned to New Zealand in 1958 to head the heart unit at Greenlane, a role he was to fill for 30 years.

Right: Sir Brian Barratt-Boyes wears magnifying glasses for a delicate repair inside a tiny heart, 1979.

If you contracted rheumatic fever or developed some other problem with a heart valve before 1958, the chances of you surviving were virtually nil. The valves are the muscles and sinews that control the flow of blood into and out of the heart, which in turn is in control of the circulation of blood around your body, so any major problem with these valves was fatal – and untreatable. That was until Kiwi surgeon Brian Barratt-Boyes (b.1924, d. 2006) pioneered radical heart treatments at Greenlane Hospital in Auckland.

Sir Brian – he was knighted in 1971 – was born and educated in New Zealand, going overseas after gaining his MD to further his training in the US. He returned to New Zealand in 1958 to head the heart unit at Greenlane, a role he was to fill for 30 years. In the late 1950s the practice of heart surgery was brand new – Christiaan Barnard and others had pioneered techniques just a few years earlier –

Above: A commemorative stamp from the mid-1990s.

Barratt-Boyes' work during the late 1950s and 1960s was among the finest done by a New Zealander, and certainly made New Zealand a world-class centre for medical research.

and Barratt-Boyes set about creating a world-class facility at Greenlane to study it.

Practising on sheep and cadavers, Barratt-Boyes and his team worked out a way to replace faulty heart valves in humans with those from donors, taking the heart valves from dead bodies soon after death and storing them (in a fridge), for up to four weeks. They reasoned – and it is still considered the case today – that using transplanted heart valves would be more successful, and more natural, than creating valves out of synthetic materials.

Barratt-Boyes, recalling the first time he performed the operation, described himself as 'pretty tense', but considered the operation went well. For many years he kept in touch with his first guinea pig/ patient, who was at the time only 16 years old. Barratt-Boyes was also one of the first surgeons to implant pacemakers, which were manufactured in the university workshop before they became commercially available.

Another major breakthrough was Barratt-Boyes' work in 1968 on heart operations for infants. The problem with operations on infants and small children is that the heart–lung machines used to replace the normal heart functions during operations were too powerful for use on children, and when you also take into account the small size of a baby's heart, it makes for a very tricky procedure. Barratt-Boyes pioneered the use of 'profound hypothermia' on children – which sounds a bit like science fiction. It involves cooling the infant's body temperature down from the normal 37°C to about 15°C, at which point the blood stops circulating and the heart stops beating. Once the patient is in this state, surgery can begin. The patient can stay this way for up to 45 minutes.

Without a doubt, Barratt-Boyes' work during the late 1950s and 1960s was among the finest done by a New Zealander, and certainly – for a time at least – made New Zealand a world-class centre for medical research. His work brought surgeons to Greenlane from all around the globe, and turned the hospital into one of the foremost research institutes in the world. In turn, Barratt-Boyes travelled internationally for many years learning and lecturing.

Barratt-Boyes himself suffered from heart disease, a condition which only became public when a colleague performed a coronary artery bypass on him in the mid-1970s. In all he had four major heart surgeries himself, and he died from complications following a procedure in the US in 2006. The operation he was undergoing was the replacement of two heart valves – the technique he himself pioneered.

STRmix

ESR GETS ALL CSI ON DNA
AND SELLS STRmix TO THE USA

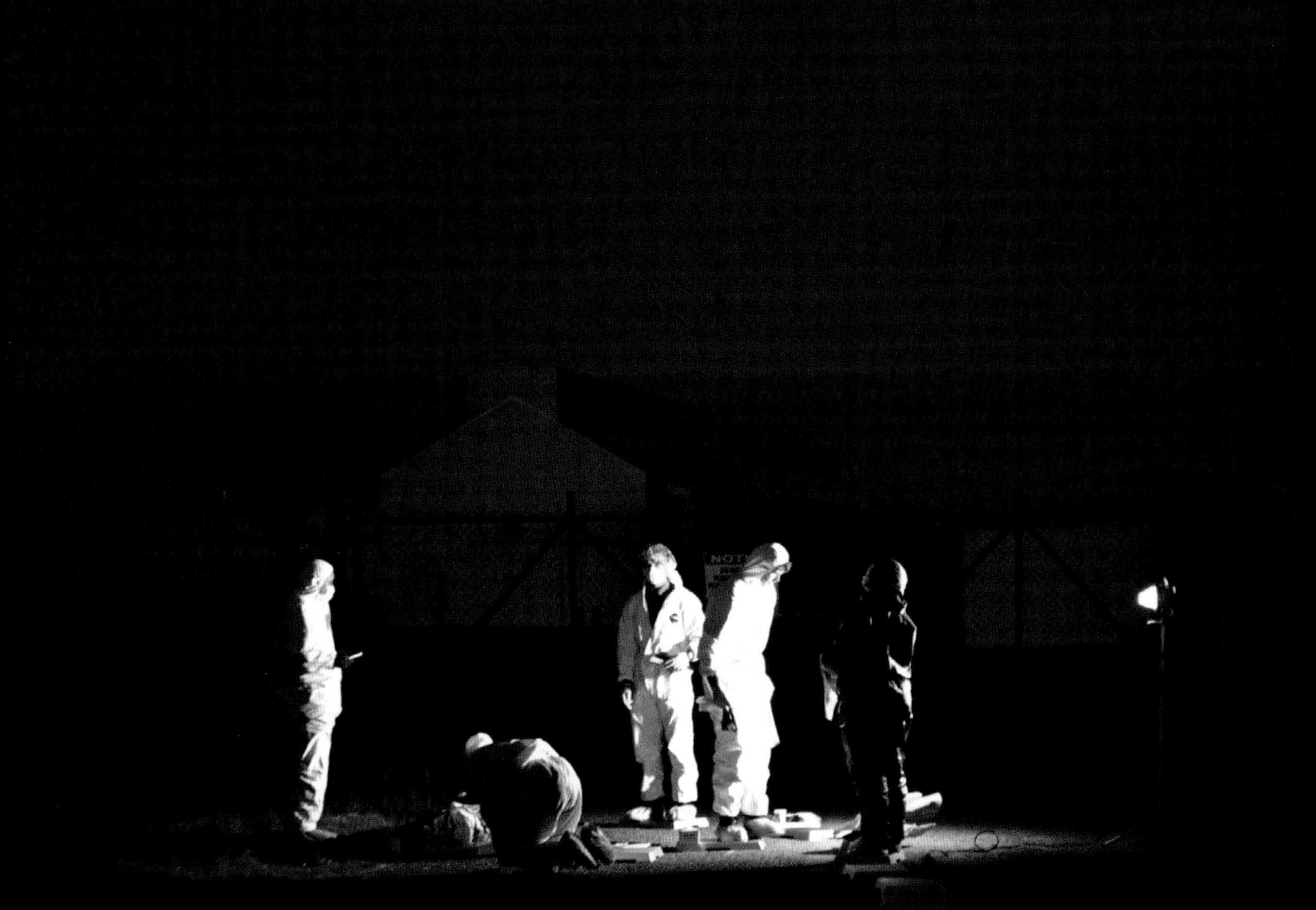

Above: Police forensics experts gathering evidence.

The liquid Australian sun beats down on a detective crouching on a kitchen floor. Around him the crime scene is being photographed and taped off by men in white overalls. He looks up at his assistant and points to a minuscule drop of blood on the lino. 'Take that for testing, darl,' he drawls. Two days later the results show more than one person's blood in the sample, which means the DNA can't be identified. Streuth!

The inventions in this book have sometimes been daring, large, loud, fast and exciting, but there is none more Hollywood than STRmix – you might think the name could use a bit more glamour, but it's pronounced 'star mix' so there you go. This invention is straight out of an episode of *CSI* – a future episode, that is.

Up until a couple of years ago, when two or more people's DNA was mixed together in a single sample recovered by the police there was trouble. Not only has something criminal gone on – or something very naughty at least – but also the traditional methods of interpreting and matching DNA could not resolve the sample and match it with records

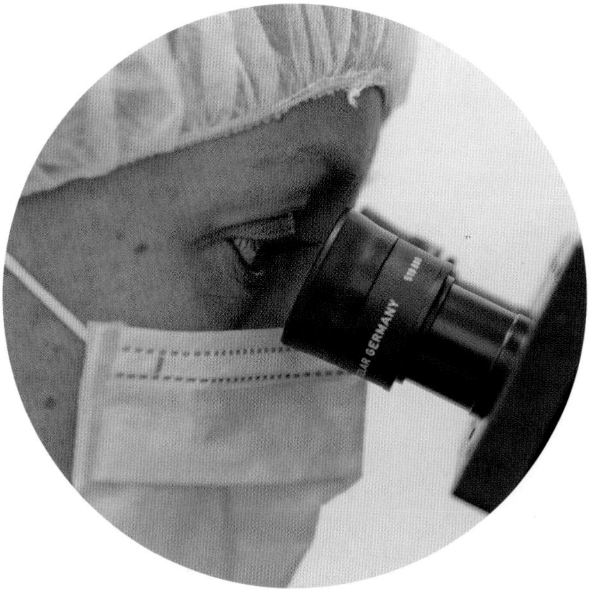

2009

In 2009 the police chief of Victoria in Australia banned DNA evidence in court after errors interpreting it were discovered. To remedy the crisis, Australian and New Zealand scientists got together to try and improve and standardise the way testing was done.

held on a database. Any vague hints given by the interpretation would be too sketchy to stand up in court. The DNA sample was useless and the perps would get away with it.

ESR isn't quite *CSI* or the FBI, but their research and testing in all sorts of science underpins New Zealand's health and justice systems. Environmental Science and Research scientists, based in their multi-storey building on the northern slopes of Auckland's Mt Albert, are the forensic DNA analysts for the New Zealand Police. ESR scientist Jo-Anne Bright calls the situation where no DNA match can be made a 'stop', and prior to 2013 ESR had to report many DNA investigations as stops, also bringing a stop to the police investigations, or at least hampering them greatly. As Bright says, 'It was quite hard to interpret in the olden days – the olden days being last year!'

In 2009 the police chief of Victoria in Australia banned DNA evidence in court after errors were discovered. To remedy the crisis, Australian and New Zealand scientists got together to try and improve and standardise the way testing was done. Bright and Dr John Buckleton as well as an Australian, Dr Duncan Taylor, were charged with the task (but for the purposes of this book we'll probably just play down the Australians' involvement a bit).

Just two years later and ESR had a product that did more than fix the existing problem – it's now revolutionising DNA interpretation all over the world. It's complicated science but Bright describes their breakthrough like this: 'We used some standard maths and some clever biology, we removed the subjectivity that was previously around DNA matching and combined that with some software that a biologist could represent in court.'

The Australian scene fancifully described above is real – it was one of the first of STRmix's many successes so far. Detectives found blood at the home of a suspect in a multiple homicide, but the sample proved to contain blood from more than one 'vic'. In the past it would have been useless, but STRmix unravelled the DNA, and placed the victims at the crime scene. They put the guy in jail.

And it's not only good news for solving future crime – the method has already successfully solved past crimes that couldn't be processed with the technology of the 'olden days'. The number of crimes like this around the world that could now be solved is immense.

STRmix is now the Australasian standard method. The US Army has already purchased the technology for use by the military police and in anti-terrorism investigations, and the FBI and ATF (Bureau of Alcohol, Tobacco, Firearms and Explosives) are in talks to buy it as well. Those are the big three federal crime agencies. If they all adopt STRmix, it's likely state and county labs all over the US, and the world, will follow suit. How big could this be? Bright laughs nervously. 'It could be bigger than ESR, which is a bit daunting, because there are only three of us!'

So next time you watch *CSI Miami* or *CSI Las Vegas*, just remember they might be glam and slick, but the investigations that get real results are all on CSI Mt Albert.

Humidifier

IT'S JUST A LOT OF HOT AIR

40

Sleep apnoea sufferers may stop breathing for up to 40 seconds.

Right: The original humidifier made from a domestic Agee preserving jar. Far right: Dave O'Hare and an early humidifier, circa 1970.

Working at DSIR, engineer and inventor Alf Melville created the first prototype for Dr Matthew Spence at Auckland Hospital using an Agee jar borrowed from his wife.

If you lie awake at night listening to the sound of your partner snoring, you can either shove them so they roll over, or you can blow warm, moist air up their nose. While you may have used that first option, the second one may sound a little strange – but it's the clinically approved way to treat people with obstructive sleep apnoea, and the way one of New Zealand's largest and most successful companies makes money.

Sleep apnoea is a condition where an obstruction or blockage in the upper airway stops you breathing while you're sleeping. It's common for the sufferer to not even know it's happening, but they may stop breathing for 20 to 40 seconds, resulting in a poor night's sleep because the blood isn't getting the oxygen it needs. Lots of other health problems can crop up because of this condition. One way to fix this is to ensure a steady positive flow of air – and it had better be slightly warm and moist otherwise it will make the situation worse. This, in essence, is what Fisher & Paykel Healthcare's products do.

Fisher & Paykel Industries was renamed Fisher & Paykel Healthcare Corporation in 2001; F&P Appliances was spun off, and the two businesses became separate entities. Their history though, lies in the willingness of Fisher & Paykel to innovate in the late 1960s, and their tenacity in supporting a fledgling business through its initial stages.

F&P started work on commercialising a respiratory humidifier after the company decided they should try to apply their skills in manufacturing and electronics to new areas. Working at the Department of Scientific and Industrial Research (DSIR), engineer and inventor Alfred (Alf) Melville (b. 1916, d. 2006) created the first prototype for Dr Matthew Spence at Auckland Hospital using an Agee jar borrowed from his wife. Melville worked out how to construct a wire within the jar that warmed water and air, a system for airflow, and the attachments required to connect it to a patient. It was a breakthrough that the company is rightly proud of, and the original jar is still displayed in their head offices in South Auckland.

From this humble beginning has grown New Zealand's largest medical device company, and one of our largest exporters. It took time and investment from the parent company to ensure the small humidification business led by F&P engineer Dave O'Hare took off, but after years of investment and R&D, success started to come.

F&P Healthcare have stayed in speciality niche areas, with products for use in respiratory care and sleep apnoea as their core focus. Melville contributed a lot of inventions to the medical device industry. Working in cardiac care and other fields of medicine and surgery, he became a key figure in New Zealand biomedical engineering.

To keep ahead of the competition, F&P Healthcare spend more than 8 per cent of their substantial revenue in R&D, and have a clever business model where they develop a large percentage of recurring revenue from consumables – the 'razor blade' business model (once people buy your device, they are then locked into buying necessary consumables for years after). With more than $600m in revenue, over 2500 staff and selling into 120 countries, they are one of the best examples of an ingenious Kiwi spirit becoming a global success story.

4. Spitting the Dummy

PAVLOVA, FLYING MACHINES AND RAY GUNS

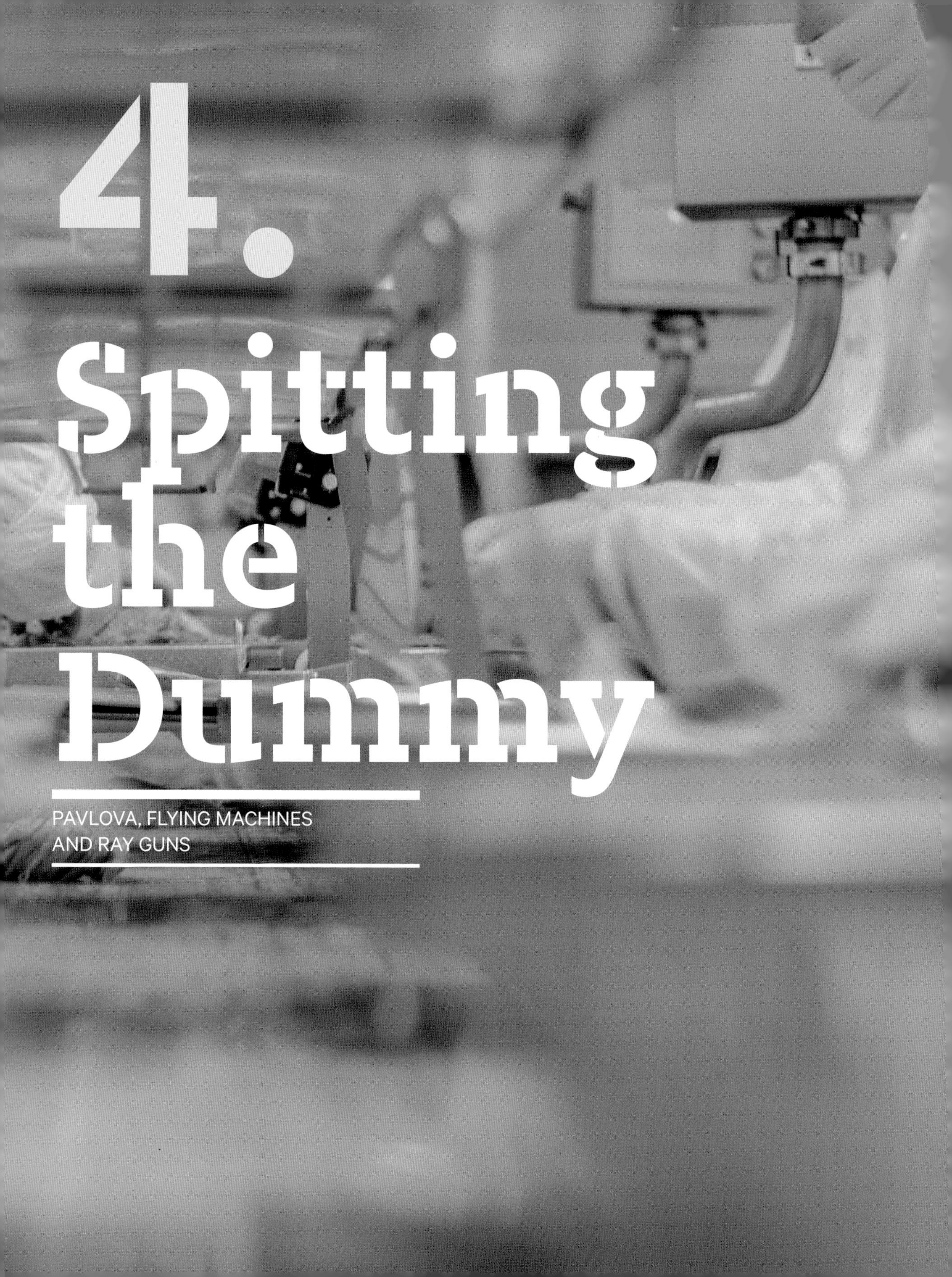

This chapter is the juicy one, full of inventions that caused major (or minor) controversies. We New Zealanders are so proud of our inventive nature that we are super quick to get very upset if somebody else claims an invention we think is ours, or if somebody tries to discredit one of our inventors or inventions. The reason these slights cut so deep is because the No. 8 Wire spirit is a big part of our national identity. If you're going to say we didn't invent pavlova you might as well stab a kea.

When we feel wronged, we 'spit the dummy' – that's Kiwi for 'throw a wobbly', which is also Kiwi for 'chuck a spazz', which is also Kiwi for 'go off your nut'…which, well, it means we get upset. In this section we explore controversial inventions and divisive ideas. Some are inventions we fight to claim, others are ideas we might prefer that others own, and some are just in themselves controversial.

The Pavlova

MERINGUE-BASED GASTRONOMIC NIRVANA

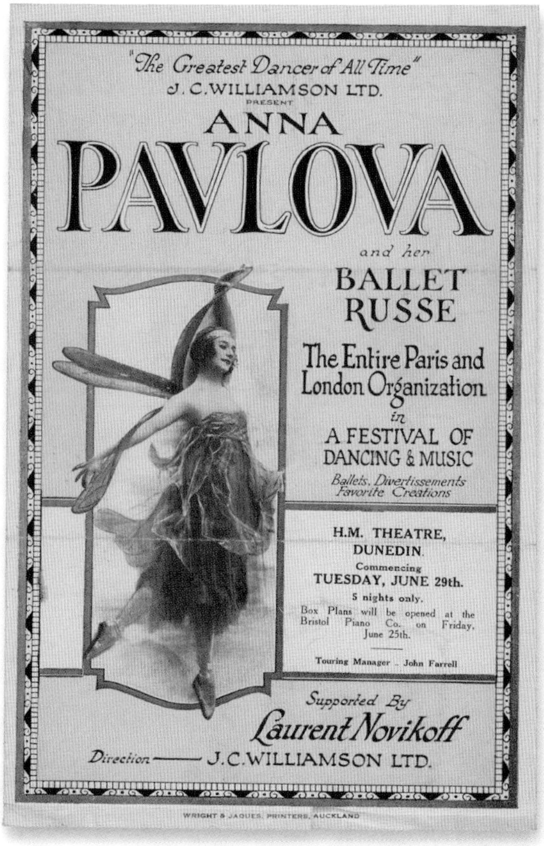

New Zealand and Australia share a friendly rivalry over many things; rugby, cricket, CER, GST... but there's one major thing that we compete over that is no joke. This is deadly serious. Come the year 2101, when the War of the Tasman is over and the nuclear cloud over Australasia has dissipated, our great grandchildren will be wandering around the ashes of New Zealand and Australia thinking, 'All this over a cream-covered meringue cake?' Because if there's one thing that would make us go to war with our trans-Tasman cousins, it's over the right to say, 'We invented the pavlova.' With the desire to finally put this thorny issue to bed and avoid any unnecessary bloodshed, we examine the evidence, impartially and without bias, and unequivocally state that it was a Kiwi who invented the pav.

This much is clear and undisputed – in 1926 a Russian ballet dancer, Anna Pavlova, visited Australia and New Zealand. She danced with such grace and lightness as to inspire the inhabitants of both countries, even the cultureless Aussies (sorry, got a little less than impartial there for a moment, won't happen again).

Another undisputed fact: There exists a cake made of meringue with a topping of cream and fruit (most properly kiwifruit, but we concede that passionfruit also works) that is also light and graceful. Said cake is popularly known as the 'pavlova'.

Now here's the thing – a) who invented the cake? b) who named it 'pavlova'? Read on and be enlightened. Be warned, though, that the twists and turns of this ancient mystery may confuse and concern, and at times look bleak and dark, but in the end, we assure you, the side of goodness and right prevails.

Read on and be enlightened – be warned, though, that the twists and turns of this ancient mystery may confuse and concern.

In the early 1930s, Herbert (Bert) Sachse was the chef at the Esplanade Hotel in Perth. In 1934 he was asked by the manager to create a new delight for her favoured afternoon teas. He laboured and experimented for a month, and then, as was tradition, he presented the results at a meeting. The cake was a meringue, covered with cream and fruit. It is said that at the meeting, the manager remarked, 'It is as light as Pavlova' – and so the new cake was named. This story constitutes the core of the Australian claim to the cake.

But don't despair, for not willing to let the matter end on this hearsay and heresy, many New Zealand researchers have searched tirelessly to restore the good name of the Kiwi cooks, and have uncovered two key pieces of evidence.

The National Library in Wellington has in its collection a cookery book, published in 1929 – a full

1927

In 1927 – eight long years earlier than Bert – the good ladies of the Terrace Congregational Church in Wellington published the second edition of 'Terrace Tested Recipes'. In it was a recipe for meringue cake, sent in by a Mrs McRae, which is exactly the same as the pavlova.

Left: A poster for Anna Pavlova's New Zealand tour, 1926.

five years before the purported cake presentation in Perth – which contains in it a recipe for 'pavlova cakes'. So it may seem that the Kiwi claim is paramount? Not really, because although the ingredients are similar, the recipe describes the process for making three dozen small meringues, not the good old pav as we know it. Damn, our claim begins to look shaky.

But not for long. In 1927 – eight long years earlier than Bert – the good ladies of the Terrace Congregational Church in Wellington published the second edition of 'Terrace Tested Recipes'. In it was a recipe for meringue cake, sent in by a Mrs McRae – blessed be her name – which is exactly the same as the pavlova. Subsequently, similar recipes were published in other magazines in the early 1930s. And then, to put the kiwifruit on the cake, Bert Sachse himself admitted in a magazine article in 1977 that his creation was really an attempt to improve earlier recipes.

So for all intents and purposes, the inventor of the cake must surely be the mysterious Mrs McRae – but what about the naming of it? It looks like the Aussies may have the jump on us – although they didn't invent the idea of naming cakes per se after Pavlova, it looks like it was they who christened the cake we know today. Interestingly, in a parallel piece of research, gastronomic historians have also recently proven that the lamington, that coconut-covered cake of controversy, was also a Kiwi invention with an Aussie name link. Originally called the 'Wellington cake', it was renamed for the English Lord Lamington, governor of Queensland, who visited New Zealand in 1895 and apparently loved the treat.

Those Aussies are always nicking our ideas… Now, Phar Lap on the other hand…

MY MUM MAUREEN DOWNS' RECIPE FOR PAVLOVA

3 egg whites
3 tbsp cold water
225g sugar
2 tsp cornflour
1 tsp vinegar
1 tsp vanilla essence

Preheat oven to 180°C. Beat egg whites until stiff then beat in water, and add sugar a little at a time until it's all used. Beat until all the sugar is dissolved. Fold in cornflour, vinegar and vanilla. Line a 20cm cake tin with baking paper or buttered greaseproof paper. Spoon in the mixture and bake for 15 minutes, then turn oven down to 120°C and bake for 45 minutes. Remove from oven and cool well away from draughts. Top with cream and your choice of fruit. Enjoy!

Light-proof Milk Contai-ner

KEEPING MILK IN THE DARK

Critics have said that lightproof milk containers create a solution for a problem that didn't exist. Dairy giant Fonterra claims that the new container they launched in 2013 was only trying to mimic the cow. After all, cows aren't see-through. Fonterra's tests indicated that 70 per cent of blind test subjects – subjects who didn't know what they were drinking, not necessarily blind people testing – preferred the taste of milk coming from their new, lightproof bottle. The bottle's three layers of lightproofing mean the contents are at less risk of being spoiled by exposure to light.

Why then did consumers nominate the new bottle for the year's 'Unfit Packaging Award'? It seems that in their hurry to innovate, Fonterra may not have realised that some consumers don't like the idea of their milk being wrapped in three layers of plastic coating, and most consumers don't like it when they can't see how much milk is left in the bottle. Ironically though, this has itself led to a new wave of innovation as consumers invent ways of seeing how much milk they have left – one genius creating a 'balsa wood ballcock' to stick out the top of the bottle. Perhaps this whole thing could have been avoided had Fonterra realised that basically, people don't like change, and unlike the new bottle, they will be able to see through any marketing gimmicks used to force it on them.

Zinc Treat-ment for Cows

THE GREAT ZINC LINK STINK

Gladys Reid (b. 1914, d. 2006) was known to the science establishment as a crackpot. 'Mad Glad' they called her, and ridiculed her publicly, causing her fellow farmers to also doubt her and laugh at her. For from 1959 Gladys insisted on proclaiming that facial excema – a dreadful disease in cows – was caused by a deficiency in zinc. She'd used her training as a dental nurse, and some careful experimenting with her own dairy herd in Te Aroha, to show that adding zinc sulphate to the water trough significantly lowered the incidence of facial eczema, even when other farms in the area suffered.

Despite this, scientists and officials refused to take her findings seriously. The Animal Health Board even noted in 1975 that zinc was 'completely useless as a form of treatment'. Reid battled on, trying to prove her case. Despite getting overseas recognition, it wasn't until 1981 that she was vindicated in New Zealand. That year farmers were advised to spray their farms with chemicals that depleted zinc in the soil. The practice caused a big outbreak of facial eczema and the Establishment was forced to back down. They admitted Reid had been right all along. She received an OBE in 1983 for her research efforts – and, one suspects, for her perseverance in the face of opposition, scepticism, sexism and most likely prejudice against people from Te Aroha.

Fetal Health

ONE OF NEW ZEALAND'S GREATEST SCIENTISTS

1

In just the first year of trials, at just one hospital, Dr Liggins had saved lives.

Sir Graham 'Mont' Liggins (b. 1926, d. 2010) is celebrated around the world because the procedure he invented has saved the lives of hundreds of thousands of babies in the last 25 years.

We'd like to make New Zealand look more sophisticated by saying that this invention is all about medical science and has nothing to do with sheep and farms. Not quite true, unfortunately. It is all about science but it was also born on a farm.

An obstetrician in the 1960s at National Women's Hospital in Greenlane, Auckland, Liggins was carrying out experiments in his 'spare time' and was trying to figure out what causes some women to give birth prematurely. He remembered a farmer friend years before who had observed that ewes who'd been worried by dogs always give birth prematurely. Liggins wondered if the dog attacks had caused the release of the stress-hormone cortisol, thus triggering the births. He began tests by administering the hormone to pregnant ewes.

Sure enough, lambs were born prematurely. Then came the surprise and the discovery. He had expected the premature lambs to die – like many premature human babies do – because their lungs were underdeveloped. They didn't. Many scientists would have just thought, 'That's curious,' and moved on. Not Liggins. His genius was that he recognised the significance of that unexpected result.

'To my surprise this lamb was still breathing. It had no right to be. It was so premature that its lungs should have been just like liver, and quite uninflatable… when we came to do the autopsy the lungs were partly inflated and this was absolutely surprising. So, weighing this up, I postulated that the cortisol had accelerated the maturation of enzymes in the lung that caused accelerated maturation.'

Infant respiratory distress syndrome (RDS) is one of the leading baby-killers in the world. Premature babies' lungs lack 'surfactant' to help them inflate and deflate easily. In 1969, realising he was on the brink of a treatment for RDS, Liggins organised randomised controlled trials on humans at National Women's.

Over one year, 282 women at risk of premature delivery were studied. Half were treated with steroids, the other half weren't. The results were incredible. A quarter of babies born to the group not given steroids had RDS, and most of those died. Of the group given the steroids, only 4 per cent of the babies developed RDS. In just the first year of trials, at just one hospital, Dr Liggins had saved lives. Imagine how much he must have wished he could have treated them all.

To the great shame of the medical profession, it took three years for a journal to accept the paper for publication, and then almost 20 years for the treatment Liggins invented to be implemented worldwide. It seems that, as an obstetrician, Liggins was encroaching on the turf of the pediatricians, and as an antipodean it was hard to gain credence in the northern hemisphere. Drug companies may also have been complicit in this great tragedy because the drugs to prolong pregnancy are expensive whereas a simple steroid costs only a few cents.

The good news is that the story of medicine's failure to respond to the studies and the resulting unnecessary deaths became a famous and much-needed kick in the pants for the Establishment.

By the early 1990s Liggins' treatment was finally standard world practice and in 1991 Liggins was given a well-deserved knighthood. The world owes him (and all those lambs) a major debt of gratitude.

Powered Human Flight

UP, UP AND A LITTLE WAY

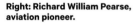

Right: Richard William Pearse, aviation pioneer.

Our ability to jump into a metal tube and fly into the sunset is so commonplace today that we take it for granted. But for most of the history of mankind we've only been able to dream of flying through the air as the birds do. A flying machine has been the goal of many inventors – including, most notably, Leonardo da Vinci, who must rate as history's best theoriser but worst actualiser. Foreign, he was.

The invention of a flying machine relied first on the discovery of suitable scientific principles to base it on. It's all very well watching birds and guessing that wings might be important in flying, but it took Daniel Bernoulli's work on aerodynamics and fluid dynamics to get enough science together to use in aircraft design. Indeed, it wasn't until the beginning of the twentieth century that serious work could be done on the problem of powered flight. Simultaneously around the world a number of inventors were working on the problem; the ones that concern us are the Wright brothers of America, and one Richard William Pearse (b. 1877, d. 1953) of New Zealand.

Thanks to their subsequent successes, quite a bit is known about the Wright brothers and their supposedly, and somewhat literally, ground-breaking achievements. They were born in Indiana and Ohio in the US, in the late nineteenth century. They were bicycle mechanics who were fascinated with flying, and worked together over a period of years perfecting gliders and playing with engines until finally, on 17 December 1903, they attempted their first (public) powered flight. Wilbur tried first, but in an inauspicious start, the engine stalled at take-off. They tossed a coin to see who would try next, an honour which Orville won. At 10.35 am he made a heavier-than-air, machine-powered flight, lasting just 12 seconds and covering a distance of 36.5 metres – a world first... or was it?

In contrast, not too much is known about Pearse. He also was born in the late nineteenth century, and worked for a while as a bicycle mechanic – indeed, Pearse's first patent relates to a new type of bicycle he invented. But, like the Wrights, it was flying that was Pearse's passion. In 1902 he built a prototype plane and engine, and then later attempted a powered flight. His too was an inauspicious start – after a flight of around 140 metres he crash-landed into a gorse bush on his Waitohi property in South Canterbury.

Above: A model of Pearse's original 'plane' displayed in the South Canterbury Museum.

The key question is 'when?'. When was this attempted flight? No records were kept of the event, but according to circumstantial eyewitness testimony, the date was most likely to be 31 March 1903. That would pre-date the Wrights' flight by eight months!

The controversy here really is twofold – whether in fact the attempted flight was before or after the Wright brothers, and whether such a short 'flight' really counts at all anyway. Even Pearse himself was convinced that a 'flight' meant a powered take-off followed by a 'sustained and controlled flight', which he didn't feel he had achieved. Others, however, would state that his attempt was of a length that could not be done without the help of an engine, and therefore was powered human flight.

To short-circuit the arguments, let's agree that it doesn't really matter. Pearse was a great inventor in his own (w)right, and he created what looked very much like what we consider a plane today – without a huge staff of engineers or the benefits enjoyed by – as he put it – 'men who had factories at their back'. He was a true pioneer innovator, testing the limits of the known. As he said, 'It is impossible to assign any invention to one man, as all inventions are the products of many minds.'

Pearse went on to create a number of interesting aviation-related inventions, including the aileron.

140

His first attempt was inauspicious. After a flight of around 140 metres he crash-landed into a gorse bush on his Waiotahi property in South Canterbury.

Unfortunately, he never patented it so didn't really benefit from his idea. During World War II he invented a plane that could take off and land vertically – presaging the Harrier Jump Jet and similar planes. He also created some ingenious farm equipment, including an automatic potato planter with mechanical arms.

Unfortunately, in his own time Pearse was misunderstood and often maligned. Indeed, he was given nicknames like 'Mad Pearse' and 'Bamboo Dick' (because his name was Richard and he made stuff from bamboo, not a physiological jibe).

Pearse died alone in a mental institution in 1953. His spirit of inventiveness lives on, and so does his name, which is honoured by Timaru Airport and on stamps and dozens of streets.

The Eight-hour working Day

WHAT A WAY TO MAKE A LIVING

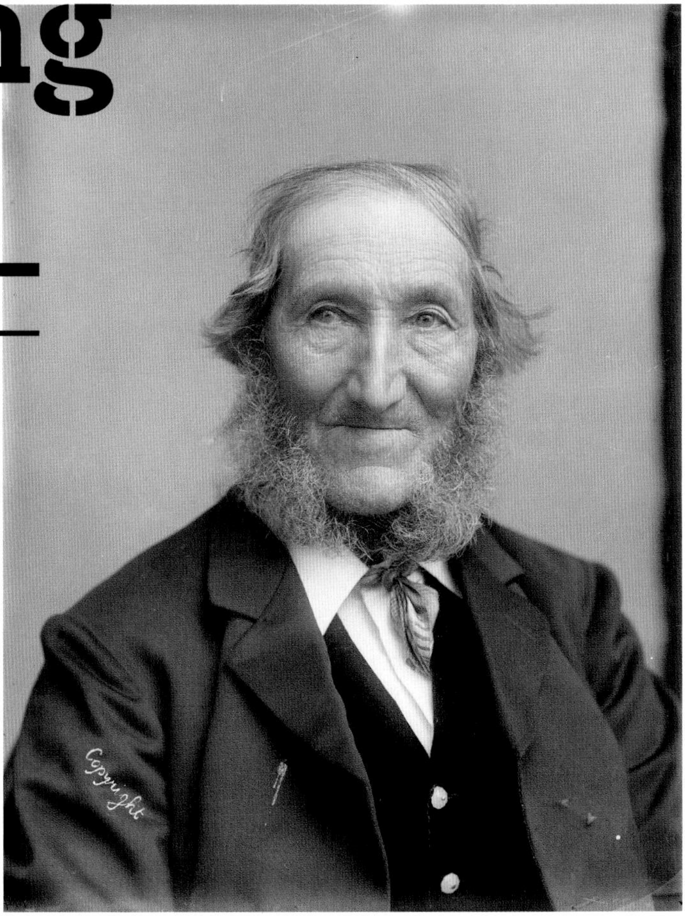

12–14

Parnell was born in England and trained as a carpenter, but working in London at the time meant 12- to 14-hour days, low wages and poor working conditions.

Right: Samuel Parnell, rabble rouser and champion of the people.

Given this innovation happened on or about 8 February 1840, just two days after the signing of the Treaty of Waitangi, it may also count as the first New Zealand invention – although of course Māori had a fair few before this date. Certainly it was a major contribution to the conditions of workers everywhere in the world – and it started on the Petone foreshore.

Carpenters, or 'chippies', have always been a stroppy lot. They've been trouble since that incident in the Near East in about (AD 30) where a carpenter went up against the Roman Empire and started a pretty big religious movement. Samuel Duncan Parnell (b. 1810, d. 1890) was no different. Parnell was born in England and trained as a carpenter, but working in London at the time meant 12- to 14-hour days, low wages and poor working conditions. Newlywed, he decided to emigrate to New Zealand in 1839. For the price of just £126 he could get passage to the colony, 100 acres (40 hectares) of land in the country and an acre in the middle of what is now downtown Wellington, but at the time was called Port Nicholson.

Parnell landed on 8 February 1840 and soon after was approached by one of his fellow passengers, shipping agent George Hunter, who asked him to build a new store for him. Parnell agreed, with one condition – he'd only work eight hours per day. 'There are,' he famously argued, 'twenty-four hours per day given us; eight of these should be for work, eight for sleep, and the remaining eight for recreation and in which for men to do what little things they want for themselves. I am ready to start tomorrow morning at eight o'clock, but it must be on these terms or none at all.'

Hunter was not giving in that easily though. 'You know Mr Parnell,' he countered, 'that in London the bell rang at six o'clock, and if a man was not there ready to turn to he lost a quarter of a day.'

'We're not in London,' replied Parnell and with that he turned his back and started to walk away. Hunter, knowing when he was beaten – and realising that Parnell's skills were in short supply in the growing colony – called him back and agreed the terms. As Parnell wrote later, 'The first strike for eight hours a day the world has ever seen was settled on the spot.' New Zealand also became the first country where the eight-hour workday came into usage.

Parnell kept up the rabble-rousing, meeting incoming ships as they arrived and talking to the workmen, encouraging them to follow the same practice. Bosses who tried to revert to the 'old country' ways of longer hours were soon discouraged, and in October 1840, a meeting of Wellington carpenters at a pub on Lambton Quay pledged 'to maintain the eight-hour working day, and that anyone offending should be ducked into the harbour'. No record is kept of how many people were actually ducked into the harbour but the movement certainly gained momentum. It was also the beginning of the 'business lunch' on Lambton Quay.

The eight-hour workday caught on with other workers and trades, and Parnell claimed victory for the movement when a road-workers' strike in 1841 secured their right to work an eight-hour day while building the Hutt road. In other parts of New Zealand, the sanctity of the working day took a little longer and required other acts of individual leadership – for example in Otago, where Samuel Shaw (b. 1819, d. ?) organised one of the first mass rallies against unwilling employers in 1849, and became a leader in the growing labour movement. Incidentally, Shaw may have also started another New Zealand hobby, complaining about the price of housing, when he wrote that in New Zealand, 'While wages were not high, the price of land was too high.'

'There are twenty-four hours per day given us; eight of these should be for work, eight for sleep, and the remaining eight for recreation.'

The labour movement was now under way, and Parnell's contribution to it was significant. In Auckland they named one of the new boroughs after him in 1877, and in 1890, when new New Zealanders wanted to look back and celebrate the achievements of the first 50 years of the young country, they used the occasion to honour Parnell's efforts. Parnell – by this stage an 80-year-old man, still living in Wellington – was put at the head of the first Labour Day parade on 28 October 1890, where he was cheered mightily and speeches were given in his honour. Parnell replied that he was so happy that day, 'because the seed sown so many years ago is bearing such abundant fruit and the chord struck at Petone fifty years ago is vibrating round the world, and I hope I shall live to see eight hours a day as a day's work universally acknowledged and become the law of every nation of the world'.

Parnell died just a couple of months later, but in 1894 the New Zealand government again led the world, passing into statute a law that recognised the role of unions in the labour movement, and creating the world's first system of arbitration for labour disputes. The legacy of activists and leaders like Shaw, and especially Parnell, was not just the eight-hour working day, but what it came to symbolise – a fair system for all.

Spreadable Butter

DECIDING THE MARGARINE VS BUTTER CONTROVERSY

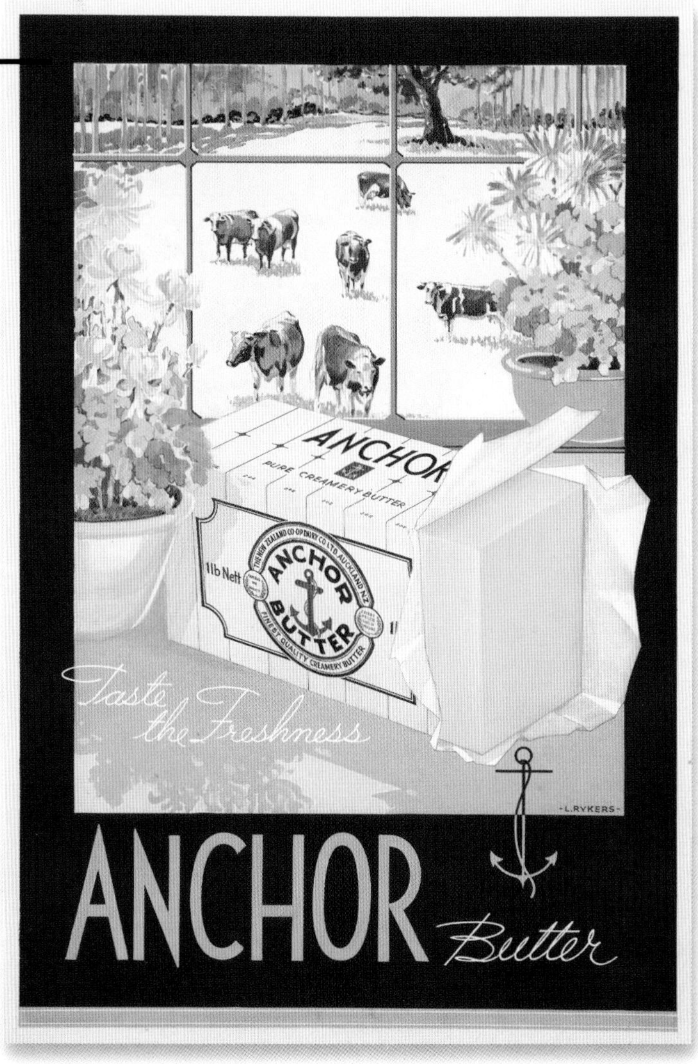

1990

The world's first spreadable butter oozed off the production lines.

Right and far right: Two of New Zealand's best loved butter brands.

Not wanting to display any bias or impartiality in the margarine vs butter debate, I won't state which side I'm on, but I think it's true to say that those bloody marg lovers have always had the killer argument. No, not cholesterol, not price, not even, laughably, taste. The killer argument for the marg camp has always been, 'Yes, but can you spread it on your toast?' And of course, they've been right. In the past it was necessary to keep the butter out of the fridge, or to bung it in the hot water cupboard for a while just to get it to a spreadable consistency. But now, thanks to the New Zealand Dairy Research

Institute, you can wipe, or rather spread, that self-righteous, polyunsaturated grin off their faces.

Dr Robert Norris and David Illingworth and the team at the NZDRI started working on the 'spreadable butter' project in the 1970s. Part of their impetus was a change to regulations which meant the English were no longer guaranteed to buy our butter – they had to want to buy it. To become viable, it had to still be butter, so the researchers had to work out a way to use the same raw materials as 'normal' butter – cream and salt – to make a spreadable butter. They determined that the secret is not what you put in; it's what you take out.

Dairy products contain a large number of different types of – and I hesitate to use this phrase for fear of giving ammunition to the margarine supporters, but it *is* the chemically correct term – fatty acids. It was discovered that some of these fatty acids, in a form known as a 'triglycerides', have a melting point in the same range as the difference in temperature between the fridge and the room. In other words, when butter is in the fridge these triglycerides are frozen, making the butter hard, but if you leave it out of the fridge they melt, making the butter soft. The NZDRI team correctly surmised that if they could somehow remove these particular triglycerides, leaving everything else intact, then the butter should be as soft in the fridge as it is at room temperature.

Doing it on a small scale wasn't too much of a problem. At the scale needed for commercial production it was a bit more of a challenge, but the team managed it, and in 1990 the world's first spreadable butter oozed off the production lines – but not onto *our* supermarket shelves. It seems the problem of hard butter was not initially considered important enough in New Zealand to warrant selling the new product here. Apparently New Zealand is one of the few countries in the world where fridges have butter conditioners in them, keeping the butter slightly warmer than in the rest of the fridge. Instead, for the first six or seven years of its creamy-textured existence, all our spreadable butter went offshore.

Apparently New Zealand is one of the few countries in the world where fridges have butter conditioners in them, keeping the butter slightly warmer than in the rest of the fridge.

In England it was an immediate hit. In 1998 spreadable butter was the fastest growing grocery product in England, with sales increasing by 37 per cent on the previous year.

But with success comes jealousy, and it came in 1996 in the form of a ban on the import of New Zealand spreadable butter into the European Union, with those – obviously marg-loving – EU officials claiming that spreadable butter did not meet the strict criteria for dairy products, as it wasn't 'manufactured directly from milk or cream'. This was pure hogwash, and the decision was overturned by the World Trade Organisation in 1999.

Spreadable butter was eventually released in New Zealand too, and became as big a hit here as it was overseas. The team at the NZDRI are to thank for the ease with which we can now butter our toast in the mornings, and the vanquishing of the margarine foe.

Of course the guys who make the butter conditioners are ropable.

The Wide-toothed Shearing Comb

NEW ZEALANDERS BEAT THE AUSSIES AT THEIR OWN GAME

1950

A new kind of comb was invented in New Zealand.

Right: Godfrey Bowen, our world-record-breaking shearer, and one of his mates.

New Zealand and Australia both have a strong tradition of sheep-shearing that is part of the defining mythology of our two countries. Shearing heroes on both sides of the Tasman have been among the greatest heroes of all. New Zealand's Godfrey Bowen (b. 1922, d. 1994) broke shearing records all over the world in the 1950s and became an international celebrity – as big a legend as Hillary. Bowen once sheared 456 ewes in nine hours (just over one minute per sheep!). He pioneered the technique, still used today, of stretching the skin on the sheep with the non-shearing hand in order to get a more even fleece. Bowen was so huge the Soviet Union honoured him as a friend of the proletariat worker. A British newspaper once compared Bowen's shearing with the 'grace' of Nureyev's dancing – such was the importance of sheep-shearing in that era. The reason we labour this point is so that your post-millennial mind can comprehend the following drama.

For in that same decade a controversy brewed that drove a wedge between our two countries in a way that has only been rivalled by a comparatively trivial underarm bowling incident.

It all began in 1898 with the development in Australia of the Wolseley shearing machine. A few years later the machine was good enough to replace hand shears in sheds all over both countries. In the beginning both countries used a simple 10-tooth comb. Twenty years later and New Zealanders were using a 13-tooth comb from the UK. It suited our conditions and although it was not taken up by Australian shearers, this was not the 13-tooth comb that caused the controversy – it was another Kiwi innovation.

In the early 1950s a new kind of comb was invented in New Zealand. The 'concave' comb was a 13-tooth comb, with some of the outside teeth bent slightly inwards. At first only one tooth was bent (now three or four teeth form the concavity), but this slight modification was a breakthrough in shearing efficiency.

Whereas in the early days of shearing, Australians came to New Zealand in droves, adding an 'o' to the end of every word – giving us 'sheepos' who kept the pens full and 'fleecos' who gathered the shorn wool – in the 1970s and 1980s New Zealanders began to go to Australia in search of work. Shearers were the foundation and the basis of the whole Australian union movement, which was gaining enormous power at this time. The shearing union was very conservative and the presence of New Zealand shearers was a big threat to the solidarity of the workers' movement. New Zealand shearers were willing to work longer hours for less pay, in worse conditions. And they had a different kind of comb – a better one. With the concave comb Kiwi shearers could shear half as many sheep again in the same amount of time as their Aussie counterparts.

Above: Bowen explaining his techniques to the new Queen of England and Prince Philip. Note the ceremonial black singlet.

With the concave comb Kiwi shearers could shear half as many sheep again in the same amount of time as their Aussie counterparts.

The whole Australian shearing rate was based on the old 10-toothed comb. The concave comb – called the 'wide comb' by the Australians – became the scapegoat for all the grievances Aussie shearers had against the New Zealand 'scabs' (or should that be 'scabbos'?). New Zealand shearers' willingness to work harder, combined with their new comb, meant they were an attractive proposition to Australian sheep-station bosses, who hired them to weaken the grip the unions were gaining. The concave comb was banned by the unions and there was a 10-week-long national strike by shearers in 1983.

When the furore died down, progress was the winner – finally the courts in Australia had to step in and allow the use of the wide comb. This genuine product of Kiwi genius is now the standard comb all over the world and the controversy is all but forgotten. Well, it's at least as forgotten as the underarm bowling incident.

The Death Ray

MYSTERY! SUSPENSE! INTRIGUE! OR A BIG HOAX?

6

For six months the inventor works away, protected from intrusion, and at the expense of the New Zealand government. Word has it he is at work on a weapon called the 'Death Ray'.

Below: Somes Island in Wellington Harbour, where Penny was interned in 1935.

It is 1935, between the wars. On Somes Island in Wellington Harbour, the old hospital buildings have been turned into a laboratory where an Auckland inventor called Victor Penny (b. 1900, d. 1948?) is working on an invention. Day and night, the buildings are guarded by four soldiers with rifles and bayonets. Nobody is allowed to see Mr Penny. If anyone approaches without identifying themselves, they are to be shot. For six months the inventor works away, protected from intrusion, and at the expense of the New Zealand government. Word has it he is at work on a weapon called the 'Death Ray'.

There are really two stories to tell here. The first is the story that the public of New Zealand learned at the time from newspaper reports and word-of-mouth. Because the events were surrounded by secrecy, and because of a lack of understanding of the technology Penny was working on, this story is full of misunderstandings and sensationalism. The second story is the truth behind the events, which may never fully be known. Victor Penny was a very serious man and intensely patriotic. He was sworn to secrecy over the affair, and took to his grave a good deal of these facts.

The story as the public learned it started in the early months of 1935. A Takapuna garage attendant, Victor Penny – already known for inventing a new kind of microphone – let it be known that he was being hounded by unknown persons over an invention he was working on. The papers of the day said he was developing what came to be known as a 'Death Ray'. Rumours and speculation surrounded his work. It was said he had successfully blown up explosives buried beneath the ground from quite a distance with an invisible ray that he sometimes had difficulty controlling. He was even rumoured to have blown up a flock of sheep from a distance using it. But what really got the New Zealand authorities interested was that foreign powers had apparently begun to contact Penny, offering him lucrative contracts to develop his weapon – large sums of foreign money, which Penny had rejected. He was supplied with an armed guard and an escort to and from work, but that proved insufficient.

On the night of Wednesday 19 June 1935, when the armed guard was off sick, Penny was assaulted at the Takapuna bus depot. His groans of pain brought nearby residents running to find him on the ground, his papers strewn around him. 'My papers, my papers, the War Office,' was all he could say before he collapsed. The police took extraordinary measures.

THE

INVENTION MYSTERY

VICTOR PENNY.

STILL GUARDED.

SECRECY MAINTAINED.

MONTHS ON SOMES ISLAND.

"A BIG SURPRISE SOON."

Nine months ago the "Auckland Star" drew attention to the remarkable fact that a house in Takapuna was being closely guarded by the police. This was the home of Mr. Victor Penny, garage attendant, inventor and experimenter, who on the night of June 19 last was assaulted while on duty at the North Shore Transport Company's garage.

Mr. Penny was removed to hospital, where a thick veil of official secrecy was drawn on him. He eventually left hospital, but he did not return either to his home or to his normal occupation.

What became of Mr. Penny? The "Star," after prolonged investigation, is able to say that Mr. Penny has ever since been under police and military protection while he proceeds with his experiments. From Auckland he was taken to Wellington and transferred to Somes Island, where he remained five months. He is not on Somes Island now, but he is still closely guarded.

... invention on which Mr. Penny ... working is considered so im... ...at unprecedented steps have ... only to preserve secrecy ... the possibility of any ... The nature of ... to very few ... Only one or ...

Club, one an ex-naval officer and the other an ex-artillery officer. Representations were also made to the Minister of Defence, and the matter was subsequently referred to the Commissioner of Police. Exactly who approached the Minister and Commissioner and whether the matter was referred to these two Government officers by Penny's representative or at his request is not clear. But the fact is that Penny was guarded while at work at night time and taken home safely, usually in a taxi which came from Devonport. Why ex-officers and Frontiersmen should secrete themselves in the bus depot to ensure that no harm came to the inventor, instead of the assignment being given to members of the police force is another unexplained circumstance.

An Attack in the Dark.

Penny was assaulted on the night of Wednesday, June 19, about 10.30. That night no one was playing the role of guardian angel to him. The member of the band of ex-servicemen whose duty it was to watch over the inventor that night was unable to attend owing to illness.

As soon as the assault was reported extraordinary measures were taken by the police. They were such as to make it quite plain that they were adopted for an extraordinary reason. Detectives were rushed over from the city and began a search for something in the precincts of the bus depot. No weapon was found, but it was clear that the police were looking for something else. Were certain papers and drawings missing? When found by a nearby resident, Mr. F. W. McNeill, who heard Penny's groans and cries for help, Penny was in a state of collapse. Later, when being wrapped in some blankets in the Transport Company's office, Penny kept on calling out, "The papers! My papers! The War Office!" Penny also made a rambling statement about a sawn-off shotgun.

Robbery was not the motive for the assault upon Penny. His cash from the sale of accessories and benzine was intact, as was also his personal money. What, then, prompted an unknown assailant to steal up in the darkness to an enclosure at the end of a small yard, attack the night attendant in such a way as to leave a large bruise on the lower abdomen, cause him serious con... and numerous scratches ...

VICTOR PENNY, the Takapuna inventor, photograph... inventions, a highly sensitive microphone-condenser...

More surprising still, a close watch was kept by the police over Penny's bungalow in Napier Avenue, Takapuna. For several days constables in plain clothes were on guard and paced up and down to make sure that no one except tradesmen called at the house. Every unusual visitor was challenged and no one admitted to the house except with Mrs. Penny's permission. The house was closely guarded because in one room Penny had his experimental laboratory, which was filled with radio equipment and a transmitter and other gear wor... at least £300. Not until this gear ... removed under the direction of ... Inspector Scott was the gua... drawn.

All the time Penny wa... the Auckland Hospit... institution maint... Some hinted ... requested ...

the carriage, ... in the gu... travelled as ... There sho ... whom, she ... considera... The ... inquir... hospi... tha...

On 1 July, in great secrecy, Penny was taken by train to Wellington, then by boat to Somes Island, to produce the Death Ray under military protection for New Zealand and New Zealand only.

It was 19 March 1936 when the veil of secrecy was lifted with the announcement that Penny was no longer in the employ of the government. During Penny's stay at Somes Island, there had been a change of government. The new government was no longer interested in his work because there was 'a complete lack of corroborative evidence as to the authenticity of the alleged discoveries'. The implication was that the outgoing government had spent money on a wild goose chase.

And that is as much of the story as ever came out. Penny spoke no more about what he was working on, and the story of the 'Death Ray' passed into legend. Our recent request under the Official Information Act for more information yielded nothing, because 'the information requested does not exist or cannot be found'. Smells like a big cover-up…

Some believe the truth of this story may be somewhat more interesting. Penny was a self-taught radio engineer, but because of his lack of formal university education, he had a hard time being taken seriously. A gyroscopic compass he invented to hold true north in submarines fell on deaf ears in New Zealand, so Penny sent it off to the Admiralty in London. Some months later a policeman knocked on his door with orders to seize everything to do with the compass. Penny gave it all up except the prototype which, in frustration, he destroyed. He later learned that British submarines were being fitted with a compass similar to his own.

A newspaper reporter present at a seminar given by Penny reported that he was at work on a 'Death Ray', a headline that infuriated the inventor for its ignorance. Most likely the reporter had become confused about Penny's explanation of radio control, which he was experimenting with.

But the research Penny was involved with that caused all the furore was most likely what today we call radar. Radar works by sending out a pulse of radio waves in a particular direction, then timing how long it takes for the signal to return, having bounced off an object. The distance to the object is proportional to the measured time. Penny's experiments would have failed because he could not precisely control the frequency of the radio pulses he was sending – if the frequency is allowed to vary even a little, it won't work. Because it was technologically impossible to control the frequency of a radio signal using the means Penny had at his disposal at that time, he was doomed to failure.

In January 1935 British Air Ministry personnel were considering the possibilities of a 'beam of damaging radiation', while their scientists were advising them to concentrate on using radiation for location rather than destruction. All throughout 1935 there was a worldwide race to develop and implement radar – a sort of top secret arms race. In this light, it is not surprising that the government, possibly under orders from the 'Mother Country', kept Victor Penny under wraps.

Indeed, just a few years later, New Zealander Dr Ernest Marsden (b. 1889, d. 1970), once a pupil of Lord Ernest Rutherford, was involved in radar research in New Zealand and helped the New Zealand Navy establish a radar training school during World War II. The Allies had been spooked into action by the German Navy in 1939, when it was clear they had technology that allowed them accuracy over a long range. Marsden set up the 'Radio Development Lab' and his work allowed New Zealand and the Allies to catch up rapidly to the Germans.

Denied any further assistance, Penny returned to Auckland, saying, 'I am not worried at all. I know what I've got and I have not finished my work, not by a long way.' Three years later New Zealand was called to fight in a world war. Luckily for the Germans and the Japanese, we were not armed with Victor Penny's Death Ray; unluckily for them the radar that Marsden and others worked on played a decisive role in the Battle of Britain, the sinking of the *Bismarck* and the American victories in the Pacific.

5. Travellin' On

THE BRITTEN MOTORCYCLE, THE YIKE BIKE AND THE WORLD'S FIRST PRACTICAL AMPHIBIOUS VEHICLES

More than almost anything else, New Zealand is defined by its geography – it's a long, thin country that is sparsely populated. We have one big city, a bunch of smaller cities and hundreds of towns, split across islands with mountain ranges and sizeable rivers. In some ways, this geography is big reason for innovation, but perhaps also an inhibitor.

'Studies' show that the level of innovation increases exponentially with density of population – as like minds feed off each other and spillover benefits from one area of study lead to innovation in another. Even without huge population density, as New Zealand innovation matures, we see more collective behaviour and more team-based invention. Also for much of our history we have been defined by our remoteness – the so-called 'tyranny of distance'. However, it's given us a different perspective on the world and made us a nation of travellers. From the days of Kupe, through the generations of Europeans who came to New Zealand seeking a better life, to the road legends of State Highway 1, and into the tradition of the OE (overseas experience), Kiwis have the wanderlust. No wonder that we also make inventions to help us travel. Some of them have even taught the rest of the world new ways of moving.

Hamilton Jet

DISTURBING PEACEFUL WATERWAYS SINCE 1952

300

Today Christchurch-based Hamilton Jet employs 300 people putting jets into boats for sale all over the world.

On the one hand, Bill Hamilton embodies the pure No. 8 Wire spirit of the self-taught inventor who changed the world from his workshop in the Mackenzie Basin, but the truly great thing about the achievements of the inventor of the jetboat is that he also embodies the re-wiring of the No. 8 tradition. This is because his creative genius was married to the business acumen to ensure that his legacy is not just an idea, but a company that still leads the world today.

As a boy, Charles William 'Bill' Feilden Hamilton (b 1899, d. 1978) showed the spirit of an engineer, making his own river rafts and land yachts to satisfy his need for speed. In 1912 – two years before the government completed the first hydroelectric station in New Zealand – the young Bill Hamilton constructed a dam and waterwheel to bring electricity to his family farm and his own shed.

In 1925, as a wild-haired and mad-eyed 26 year old, he was the first recorded person in the southern hemisphere to get in a car and do the ton. That alone would be enough to get him in this book. Most blokes would surely have sat back and sighed with a bottle of beer and the cricket on, dreaming of the glory days, but Hamilton wasn't ever satisfied with what he thought could be done better. He went to England and raced cars, winning half the time, and bringing back a bride. He designed and made hundreds of implements for moving and shifting dirt and making runways. He made bombs and parts for

Above: An early version of the jetboat being tested on a South Canterbury river.

guns and ammunition during the war. He made New Zealand's first ski field rope tow. He was crazy for making things. Inventions came out of the workshop at Irishman Creek like an Australian sprinter – thick and fast.

Finally, when he was 52, Hamilton turned his mind and hands to something he'd wanted to do since he was a small child – build a better way of travelling up the wild Mackenzie Country rivers of his youth. Five years later he was swamped with orders for his invention, the jetboat. His company Hamilton Jet was born and still thrives today.

The jetboat is a brilliant way to get around a river. It has high power and requires a draught of only a couple of centimetres of water, meaning rivers previously navigable only on flounder-back could now be blasted up in a big red boat full of screaming German tourists. And because a jetboat has nothing protruding from the bottom of the hull, its reduced drag makes it more efficient than a propeller-driven boat. In 1952, however, nobody in the world knew this.

In the history of boat propulsion, human beings had devised the oar, the sail, the punt, the paddle wheel and the propeller (I'm not counting the lilo with flippers). For a couple of centuries visionaries had been toying with the idea of jet propulsion on water. Benjamin Franklin drew a sketch of a jetboat in the sixteenth century, powered by hand pump. It is a major achievement that Hamilton – completely untrained in engineering, drafting, metallurgy, hydrodynamics, or even philosophy for that matter – was able to finally perfect the water jet. Of course, his secret was that he didn't do it by himself.

His first helper was Christchurch boat builder Arnold France. When Hamilton asked France to help him make a boat for shallow waters, France suggested a jet design, lending Hamilton an American pamphlet on a centrifugal-pump water jet called the Hanley Hydrojet. The Hanley design had achieved very limited success in the United States. Bill got his son Jon Hamilton to help him and they made a centrifugal pump, attaching it to the plywood hull of a 3.5-metre boat. It worked, but it managed only 9 knots – hardly a break-neck speed. In fact, not even a break-finger speed.

Undaunted, Hamilton asked a former apprentice, Alf Dick, to help. Dick came up with the idea that the spray from the back end of the jet engine should be ejected through the transom, above the water, not below it. Imagine a hose spraying out water: the hose can spray water much faster in the air than it can below the water. It was the same principle that Dick applied to the Hamilton Jet. It worked. The boat, still using the centrifugal pump, now did over 14 knots. Good, but not fast enough for the man who had been the first to do the ton.

Next, Hamilton employed an engineering graduate from the University of Canterbury. George Davison suggested replacing the centrifugal pump with an axial flow system. The major trouble with the centrifugal pump was that it created the water jet by rotating around a vertical axis. This meant that a noisy and inefficient right-angled gearbox was needed to convert the engine's horizontal-spinning driveshaft to the vertical. Axial flow is basically an Archimedes screw – a tube with a screw or propeller inside it to push the water through. A year later, with the brand new, three-stage axial flow unit on board, Hamilton was flying up the high country rivers at over 40 knots! The modern jetboat was born.

Hamilton's first commercially successful water jet was the Chinook, which went on sale in 1957. In 1960 Hamilton's son Jon drove a boat called *Kiwi* up the length of the Colorado River, through the Grand Canyon. It was the first time any boat had been able to navigate that famously difficult stretch of water and it put Hamilton Jet on the map internationally. The Colorado series of jet units that followed in 1963 were even simpler than the Chinook units and half the cost.

Bill Hamilton was knighted in 1978. His company, CWF Hamilton and Co. Ltd (now simply called HamiltonJet) had begun a journey it is still on today, pioneering one after another type of water-jet engine. The water jet was found not only to operate well in shallow water, but also to be more efficient and economical than propeller boats at speeds above 20 knots, more manoeuvrable at low speeds, smoother and quieter and safer. For this reason it has become the popular solution for all kinds of high-speed water transport – from navigating the shallow rivers around Sir Bill's workplace, to sport and recreation boats like jetsprints and jetskis, to military, search and rescue, tourism and fishing applications, all the way up to massive 60m-long ferries that carry 400 people and are powered by four jet engines each weighing up to 8 tonnes.

Today Christchurch-based HamitonJet has a manufacturing floor space of over 12,000 square metres and employs more than 300 people. Hamilton's innovation, vision and collaboration have seen over 40,000 units installed around the world, and his company has thrived for half a century as the industry leader in a niche invented by the wild Kiwi from Irishman Creek.

The Farm Bike

AS TESTED IN THE HIMALAYAS

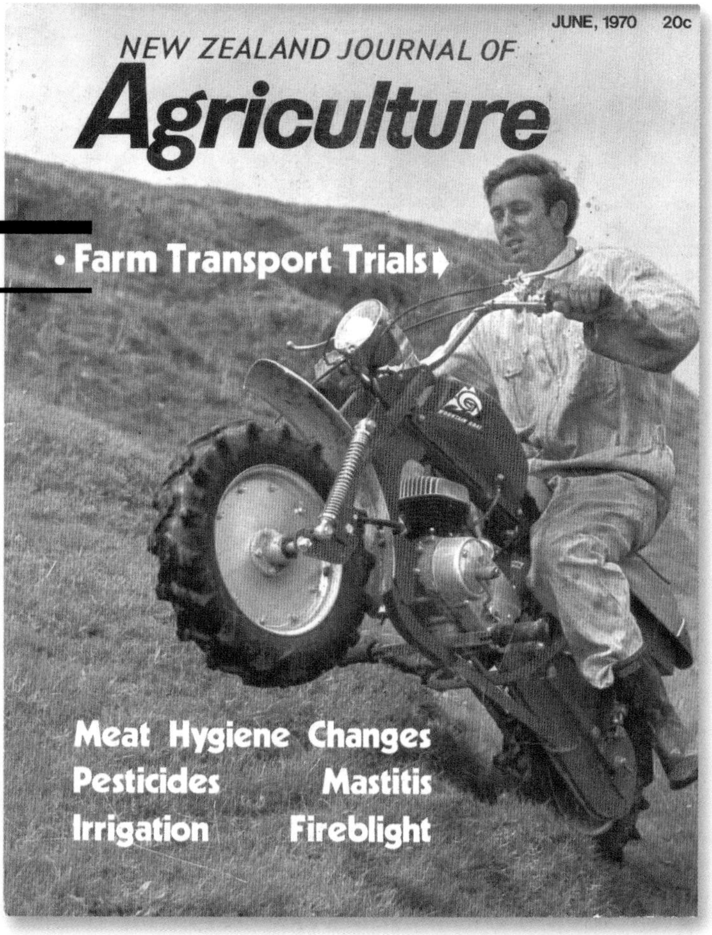

1964

Kiwi explorer Peter Mulgrew, with Sir Edmund Hillary, took the bike to the Himalayas in 1964 to try it out.

Right: The Mountain Goat in action – helmet optional, gumboots mandatory.

Cyril John (Johnny) Callender (b. 1928, d. 1978) was a New Plymouth mechanic who spotted a gap in the market. He knew from his rural contacts that the standard method of getting around on a farm – a horse – was not ideal, and that the alternative two- or four-wheeled transport options were no better. In 1963, what the cow-cockies of the 'Naki needed was a motorbike built especially for the conditions – with disc wheels to avoid sticks getting into the spokes, a slow first gear so they could follow stock slowly, enough grunt to climb the hills and light enough to throw on the back of a ute.

The Mountain Goat was born.

Not a real goat, you understand, but the name of the new farm bike that Callender created. It was locally assembled, with predominantly New Zealand parts, and named by the Kiwi explorer Peter Mulgrew, who with Sir Edmund Hillary took the bike to the Himalayas in 1964 to try it out. It performed well, and the disabled Mulgrew used it to get around the mountain camp. Callender built 120 of the original design, before competition from Japanese imports, and a bunch of government bureaucracy, stopped his endeavour. Callender's original ideas have found their way into the design of the modern farm bike, and while it's changed a bit over the years, it all started with a mountain goat on the slopes of Mt Taranaki.

The Unifoot

THE QUEEN MOTHER'S FAVOURITE NEW ZEALAND INVENTION

His Unifoot replaces the normal rubber nosing at the bottom of the walking stick with a small plate made out of plastic and rubber.

After having back surgery in 1999 Kiwi David Dell woke up one night with a flash of inspiration for a better way to make walking sticks. 'I just woke up and thought, "I'll put a foot on the end of my walking stick."' His Unifoot replaces the normal rubber nosing at the bottom of the walking stick with a small plate made out of plastic and rubber. This means the walking stick has better contact with the ground, and it can also twist to conform to the contours of the terrain. The worldwide patented Unifoot is a startlingly simple idea – at $40 it's reasonably cheap, it can be fitted to the end of pretty much any walking stick or crutch, it allows the walking stick to stand up by itself when released, and it creates four points of contact with the ground, thereby increasing the grip. Dell has now gone back to his work as a registered builder, but he sells the sticks as a sideline through his website and distributors and reckons he's sold 100,000 of them.

Let's give the British royal family the last word:

Dear Mr Dell,
Queen Elizabeth The Queen Mother has just received the special walking stick which you so kindly sent Her Majesty at the time of her 99th birthday in August.
The Queen Mother is impressed by the ingenuity and simplicity of your new invention, and it is a birthday present which Her Majesty accepts with pleasure.
Signed
(Her lady-in-waiting)

Yike Bike

THE MOST COMMON TRANSPORT DEVICE IN THE WORLD (SOON)

14

The Yike Bike is the smallest, lightest electric bike in the world. You can travel at up to 23km/h for up to 14 kilometres before the batteries need to be recharged.

Below: The Yike Bike – like a Transformer, only a bike.

It won't be long before there are 11 billion people on the planet. And a large proportion of those people are becoming more and more prosperous, and increasingly living in large cities. In fact for the first time, as we write, more than half the world's population is urban. As people become better off, one of the first things they seek is better, faster, easier and more prestigious forms of transport. And it also helps if it has a rhyming name.

Since the invention of the wheel, new modes of transport are constantly being dreamed up, tinkered with and forgotten. Few actually work, and fewer still make their way into production. Christchurch man Grant Ryan's Yike Bike is one of those few. It's like a bike, but portable and electric. It's like a unicycle but not as pointless. It's like a Segway without the humiliation.

It has been successful because it perfectly fills a niche. Imagine you live in a city (not too hard so far). You are 3 kilometres from the bus, train or subway, and the bus drops you a kilometre from work at the other end. Around the world millions and millions of people are in this situation. If you aren't keen on a long walk, you could just flag the bus and drive, or you could find something to ride to the bus stop, carry with you on the bus then ride the rest of the way to work when you get off. You'll want something you can stow easily at the office, then grab again to nip across town for a meeting. This something needs to be small, light, fast, fun, clean, safe and easily recharged. You need to be able to carry it as easily as it carries you.

You'd think that this product would have existed already, but it didn't. Electric bikes are too big and heavy to be carried onto a bus, or into a meeting. The Segway is too large and too heavy to be carried up stairs or taken on public transport. Folding bikes come close, but you have to pedal and they are mostly large and heavy too, with chains and gears complicating and dirtying things.

The Yike Bike is the smallest, lightest electric bike in the world. You can travel at up to 23km/h for up to 14 kilometres before the batteries need to be recharged. It weighs just under 15kg (or 11.2kg for the fancy carbon fibre model) and it folds up in a few seconds.

It's such a curious little vehicle that the Yike Bike company have invented a new category of bicycle to categorise it – the mini-farthing.

Grant Ryan grew up in Invercargill. He had inventiveness and the entrepreneurial instinct in his blood. His father was a chicken-plucker – Grant is not the chicken-plucker, merely his offspring – who invented, among other things, a chicken-plucking machine and a plastic bag to keep lambs warm in winter: both are sold around the world. At school when asked what he wanted to be, Ryan wrote 'inventor'. 'They told me, "You've got to put something serious." So I said, "Yeah, all right."'

Ryan set about inventing the Yike Bike very purposefully. He had already been part of four start-ups and by this fifth one he knew what he was doing. The vision was clear and ambitious – to invent the most common transport device in the world. The Segway was the inspiration and Ryan even called the Yike Bike development project Project Garlic, after Segway's codename, Project Ginger.

He started with one certainty – the front wheel needed to be about 50 centimetres – or 20 inches like an old Raleigh 20 – to allow a smooth ride. Ryan and his small team went immediately into rapid prototyping – an iterative process where you begin by making something, then you change what's wrong with it, and make it again until you reach a solution to the problem. 'We failed rapidly,' says Ryan. Twenty prototypes later the process worked. The first prototypes were pedal-powered and then they introduced electric motors. By the time they cried, 'We've found it!' they had a very new solution to human transport – the mini-farthing. A solution, Ryan is adamant, they never would have come up with on paper or in a single eureka moment.

The bike has a range of fascinating features. The electric motor at just 500g is very light and about fist-sized. But thanks to a sophisticated electronic control system that Ryan's team came up with it puts out power comparable to the best human cyclists, and gives torque even at low speeds. Sensors on the motor know where the rotor is so that torque can be applied appropriately to give better power when starting up. The same sensor control system gives anti-skid braking – a world first for a bike. The nano-lithium battery recharges itself when you brake or plug it in, and it costs only 8 cents of electricity to recharge fully. Finally it is very easy to learn to ride – just like riding a bike.

You can buy a Yike Bike today for around $4000, and Ryan encourages you to do so. He calls early adopters of technology 'unappreciated economic superheroes' as the money they inject allows a nascent technology to develop and become cheap enough to be ubiquitous. Ryan points out that it was all the people who bought the 'brick' cellphones (and were laughed at by their friends) who allowed the product to change the world as it has. You wouldn't call the Yike Bike a brick, but future versions will be lighter, simpler and most of all cheaper. Because it's essentially just a bike and a power tool, Ryan envisages them eventually selling for just a couple hundred dollars. 'Why couldn't it be the main form of transport one day?'

Compared to that dream, success has been modest so far, but the Yike Bike has still sold in the thousands. There are 75 dealers in 43 countries and the groundswell is building. The biggest barriers have been bureaucratic difficulties classifying and regulating the vehicle, and simply getting people to change their car-centred habits. 'World domination? I'll tell you in five years,' laughs Ryan.

Meanwhile Ryan has 'somebody proper' managing Yike Bike while he continues his compulsive creating. He's developed an innovation process he calls 'hunch, crunch, munch' and while Yike Bike is munching away, he's hunching and crunching with many new ideas and projects including something to be unveiled soon in the 'tourism space'. He's optimistic about New Zealand's future in innovations, citing Wynyard, Diligent, Xero and SLI (his own start-up with brother Shaun) as exciting recent developments made possible by the world-shrinking effect of the internet and the increasing availability of venture capital.

Whether or not it fulfils Ryan's vision, the Yike Bike is certainly already an iconic New Zealand invention. One of the company directors was riding his Yike Bike on the street in central Sydney when a woman called out to ask him, 'Are you from the future?' There's no record as to whether he replied, 'No, I'm from New Zealand!' but he should have. That would have been cool.

The Amphibious Boat

THE WORLD'S FIRST PRODUCTION AMPHIBIOUS BOAT

850 people in 45 countries own a Sealegs amphibious boat. That makes it not only the first commercially available land-going boat, but the most successful amphibious vehicle in the world. You've probably seen them at your local beach or boat ramp, driving audaciously into the waves, pulling up their wheels and zooming off.

We've all seen the drama that launching and retrieving a boat from a trailer can cause. The boat is half in the water. One person is struggling with a winch. Someone is lifting the kids into the boat. Uncle Robert is standing in water up to his waist holding the transom and trying to keep the boat from turning sideways. Eventually everyone scrambles in and with luck the engine starts and they motor off. You laugh at them, until you remember they've got a boat and you've got a boogie board.

Sealegs is a clever invention in more ways than one. First it is a brilliantly simple and perfectly executed technological solution. Second it is a beautiful piece of marketing in that it's a product created to fill one identified need: easier launching.

Above: The Black Beauty of Sealegs.

100

The company has now sold more than $100 million worth of Sealegs worldwide.

Left: Believe it or not, there is a boat hidden in this picture.

In 2001 he drove that boat out of his shed and into the water. 'I got some funny looks that first time,' says Bryham, 'but that hasn't changed!'

It doesn't try to be all amphibious craft to all people – it isn't a car that goes in the water, it's a normal power boat with a top speed on land of just 10km/h.

Maurice Bryham lives by the beach at Milford in Auckland. He's the founder of well-known IT company PC Direct, and 'a bit of a handyman'. Watching people struggle with their boats in the surf, he decided to solve their problem. He wasn't daunted by the fact that since the wheel and the boat had been invented, thousands had tried, and nobody had succeeded in putting them together.

Bryham did some sketches, bought an RIB – a rigid inflatable boat like the ones surf lifesavers use – and using jigs and bits of wood he converted the boat. In 2001 he drove that boat out of his shed and into the water. 'I got some funny looks that first time,' says Bryham, 'but that hasn't changed!'

The first boat had wheels that were electrically powered and controlled, 'because that's what I knew'. But over the next couple of years, Bryham made 25 prototypes and ended up with a hydraulic system. 'The hardest part was sticking to the original concept and not getting distracted with complex suggestions: my engineers and naval architects said "Let's put flaps on here or cowlings on there." I had to keep bringing it back to the original concept so it wouldn't end up designed by committee.'

In 2004 the first patent was granted, and the first boat was sold. Ten years later Sealegs boats come in 6.1m, 7.1m or 7.7m lengths, in RIB (inflatable) or D-tube (aluminium) configuration. Each option also comes in recreational form, or in professional form – for coastguard, lifesavers, rescue, police, army, harbour patrol and the like.

Two large wheels at the stern and one at the bow will drive your Sealegs out of the shed and down to the water, where you drive right in. The wheels are powered by a small motor separate to the boat's outboard engine, and pivot hydraulically up out of the water when you hit the surf.

The company has now sold more than $100 million worth of Sealegs worldwide. Speaking to Bryham as he rides a taxi to the airport to visit the Miami International Boat Show, you can hear the pride in his voice when he says it all feels quite humbling. 'It's great to see a Kiwi invention used around the world, and bring that money back to New Zealand.' It's a far cry from the day Bryham first demonstrated Sealegs at a boat show: everyone laughed at him, telling him, 'It's never going to work, it's going in the salt water you idiot, it'll be a disaster!'

How much for a Sealegs – a milestone in maritime history? They sell for between $170,000 and $240,000 – but they'll probably take a trade-in on your boogie board.

The Electronic Petrol Pump

THINGS ARE REALLY PUMPING IN THE RANGITIKEI

In 1939 Ray Williams' father agreed to loan him the money to start his own business, on one condition – that he never leave Marton. So from a small town in the Rangitikei, PEC, as his company became known, made a variety of products over the years: armaments during the war, ploughs, electric fences and petrol pumps. In the mid-1970s they began to get an inkling that a big change was on the way. The oil crisis pushed the price of petrol up so much that it was swift approaching the one dollar mark – and the old petrol pump price dials had only been built with two digits! On top of this, as the price rose higher, the price dials spun faster and more frequently, often causing the numbers to jam.

The team at PEC could see a day when mechanical petrol pumps would be a thing of the past. In true Kiwi spirit they jumped in head first. PEC bought New Zealand's first development kit for the Intel 8080 processor, and the engineers at the company taught themselves how to make the world's first electronic petrol pump – the Empec 80. Their machine got rid of the old Ferranti Packard displays – the flip-over digit ones, which worked kind of like the departures boards they still have at some airports – and in the process changed the whole dynamic of the petrol station forecourt. No longer did petrol stations need extra staff to assist their patrons with everything; the customers could now preset the amount of fuel they wanted, and new communications systems pioneered by PEC meant the data from the pump could be fed directly to the till. The look of the forecourt changed, and the term 'self-service' took on real meaning.

PEC bloomed with this new venture. During the 1970s, 1980s and into the 1990s, they led New Zealand and the world in electronic petrol pumps, and in the process grew to employ over 280 staff in Marton – easily the town's biggest employer, rivalling even the nearby mental institution.

In 1999 the Gallagher Group (see page 234) bought PEC outright and rebranded everything to the Gallagher name. Gallagher were actually after PEC's Cardax product for their security division – Sir William Gallagher said, 'We had the perimeter and this was the gateway' – but in the end they also hung on to the valuable petrol pump technology. Now the Gallagher-branded pumps account for almost all of the fuel dispensers sold in New Zealand, and about 60 per cent of the Australian market, and they are looking at expanding into South America.

The fuel division of Gallagher is now centred in Marton. In a world where microprocessors are being put into everything, the Martonites can be proud they were the first to put one in a petrol pump.

PEC bought New Zealand's first development kit for the Intel 8080 processor, and the engineers at the company taught themselves how to make the world's first electronic petrol pump – the Empec 80.

Side-Loading Tie-Down Ratchet

ONE DAY SOON YOU WILL HAVE ONE IN YOUR GARAGE

The tie-down ratchet is one of those ubiquitous pieces of gear that make the world tick. There are literally millions of them in hardware stores and garages, and millions more used by every truck driver on the planet. They are tedious to use because the straps are always much longer than you need, and you have to poke the end through the mechanism and then feed it all the way through until you take up all the slack, then the reverse when you undo them. But that's just how they work, right? Yes, right. Unless you are Barry Armour.

Barry Armour is an engineer and an inventor. Judging only by his name you might guess he invents tough things in a man's industry, and you'd be right. Barry's Dunedin-based Armour Transport Technologies had two inventions on its books – a load anchor that sits flush on a truck bed then pops up when you need to attach a chain or strap to it, and a system to make a truck 'crouch' so you can easily drive your forklift or digger off. Then Barry had his world-changing brainwave.

The Armour Side-Loading Tie-Down Ratchet has a hinged side that opens, allowing you to simply loop the strap into the mechanism wherever on its length suits you, then swing the side closed, and ratchet away. Attaching straps takes half the time – a very big deal in the time-critical transport industry. But more importantly, it's just simply a 100 per cent improvement to a global product. Why would anyone in a hardware store buy the old sort?

Armour says, 'You can probably blame my father-in-law,' who would borrow Barry's ratchet tie-downs but couldn't be bothered unthreading them, so would return them still threaded and tangled up in the back of the ute. That was in 2010, and it didn't take Barry long to sketch out the solution.

'We were saving hard to pay the patent people. They get the money first and they are *very* good at getting it.'

From there it took a year to do the CAD drawings and produce a laser-cut prototype, then another year to travel overseas, gauge interest and find a Chinese manufacturer. A third year was taken up perfecting the manufactured product with the factory: 'It ended up remarkably similar to my original sketch.' Meanwhile, says Armour, 'We were saving hard to pay the patent people. They get the money first and they are *VERY* good at getting it.' Now the product is patented in 54 countries, there's a container load in Barry's garage for New Zealand and, branded as 'GearArmour', they are on sale in big American hardware chains Lowes and Home Depot.

When asked if this is the invention that could change his life, Armour says, 'It could well be life-changing. It could be that I'm not broke all the time!'

Godward's Economiser, Hairpin and Eggbeater

PRODIGIOUS OUTPUT FROM INVERCARGILL'S FINEST

March 8, 1932. E. R. GODWARD 1,848,620

MEANS FOR VAPORIZING FUEL MIXTURES

Filed April 11, 1930

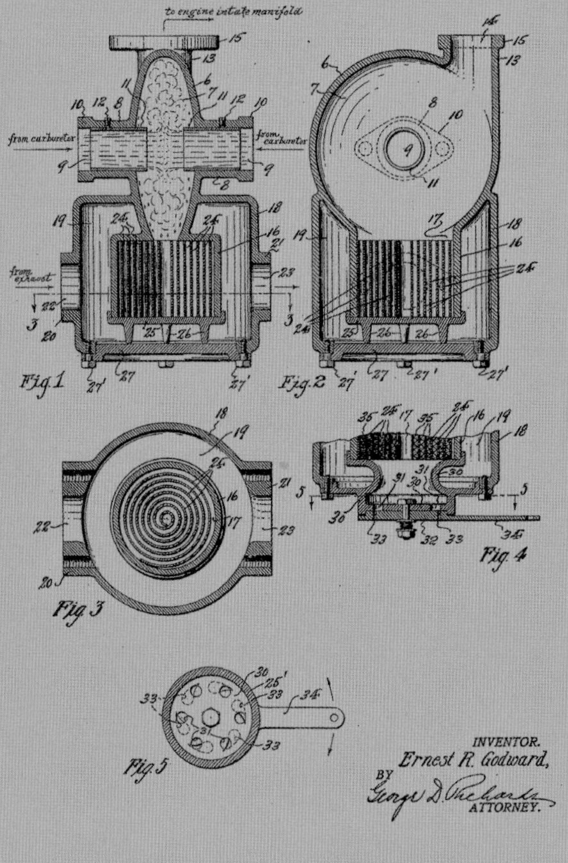

41

Godward surprised doubters by riding a motorcycle fitted with his Economiser 41 times around the Caledonian Grounds in Invercargill.

Left: Detail from Godward's patent application. Far right: Ernest Godward was an accomplished musician, painter, runner, swimmer, rower and cyclist.

This is a story about a man who needed to get out of Invercargill. Why he needed that so badly, apart from the obvious, we will never know, but Ernest Godward (b. 1869, d. 1936), champion athlete, successful painter, musician, politician, public speaker, and prodigious father must go down as one of New Zealand's most successful inventors – both for the variety and novelty of the products he developed, and the global success he achieved.

Godward had been born in England, but at the age of 12 he ran away to sea, reaching East Asia before someone noticed how young he was and sent him back. He was apprenticed as a mechanic for a few years but couldn't wait to get to sea again, this time as a legitimate steward for P&O Lines. He landed in Dunedin in 1886 and decided to call the South Island home from then on. He moved to Invercargill and set up there with a new wife, Marguerita Florence Celena Treweek, and it was a fruitful marriage – the couple went on to have 10 children.

Clearly not one to sit still though, Godward starting inventing. By 1900 he had devised a new type of eggbeater, a new post-hole borer, a hair-curler, a draught protector and many other interesting products. Real success came in 1901 when he invented a new type of hairpin, one with a spiral in it. Godward's invention became a great hit and got him out of Invercargill. He formed a company, 'Godward's Spiral Pin & New Inventions Company'– surely he should win some sort of award for the most literally named company – and taking out an international patent on it, he travelled to the United States for a year where he sold the patent for the impressive sum of £20,000 – nearly $3.5 million in today's currency. He returned to Invercargill, built a huge house – Rockhaven, which is still standing – and plotted how to get out of town again, this time for longer.

Among the amazing list of things he did in his life is making motorcycles and bicycles. While working on motorcycles, he began to fit them with a petrol economiser that he had invented. The idea is simple: before fuel can be ignited in the cylinder of an engine, it must be vaporised. The vapour is what is ignited by the spark, but if the fuel is not well vaporised then globules of fuel will condense and be sucked into the cylinder as liquid, it won't combust and it'll be wasted. Furthermore, liquid fuel unburnt in the cylinder causes more carbon monoxide pollution than is ideal. The Godward Economiser fits between the carburettor and the engine, taking the mixture from the carburettor and spinning it onto a curved, heated surface. Big drops of unvaporised fuel fall to the bottom where the surface is the hottest and are vaporised. Smaller drops land at the cooler part of the surface and are also vaporised. The mixture is delivered to the cylinder with no fuel remaining as a liquid, and is thus incredibly efficient.

The genius of the invention is that it was a simple unit that fitted into otherwise normal engines to dramatically increase their performance. In 1913 he roved to critics just how good it was, riding a motorbike fitted with the Economiser around the Caledonian Grounds in Invercargill 41 times. According to witnesses, 'The consumption of petrol was diminished to a very surprising degree.'

When Godward realised what he had, he saw his opportunity to journey overseas again. He went straight to England to capitalise on his innovation. However he was knocked back by the Poms who couldn't see past the fact that he had no formal education. Even in those times, it seems you needed qualifications. So Godward had to come home. He went to university and got his degrees, then in 1916, while the rest of New Zealand's young men were off to fight in the war, Godward was off to America. So it was that Godward Gas Generator Inc. of New York, and then the Godward Carburettor Company of London, were formed, and Godward made his real fortune.

He invented over 70 models of his economiser and became well known as an expert on the internal combustion engine.

By 1929, the US Army was fitting Godward's invention to all military vehicles. Eventually Godward Vaporisers (as they were called in the US) could be found in cars made by car companies all over the world. 580 buses in Philadelphia were fitted with the Economiser. In all, he invented over 70 models of his Economiser and became well known as an expert on the internal combustion engine.

In many ways this is the story of amazing success. Even though he was hit in the stockmarket crash of 1929, Godward's companies must have made him a very wealthy man. Perhaps more importantly, though, his inventions have had a lasting effect on the world.

Eventually the pull of the South (or perhaps the years away from his family) was too much for him. At 67 years of age, Godward boarded the steamship *Mongolia* to return to his wife and those 10 children. Sadly, though, he died before the ship reached New Zealand, collapsing after winning an onboard skipping competition. This was a man who had a drive to succeed right up to the last.

Romotow

LOZENGE-SHAPED MOBILE LUXURY

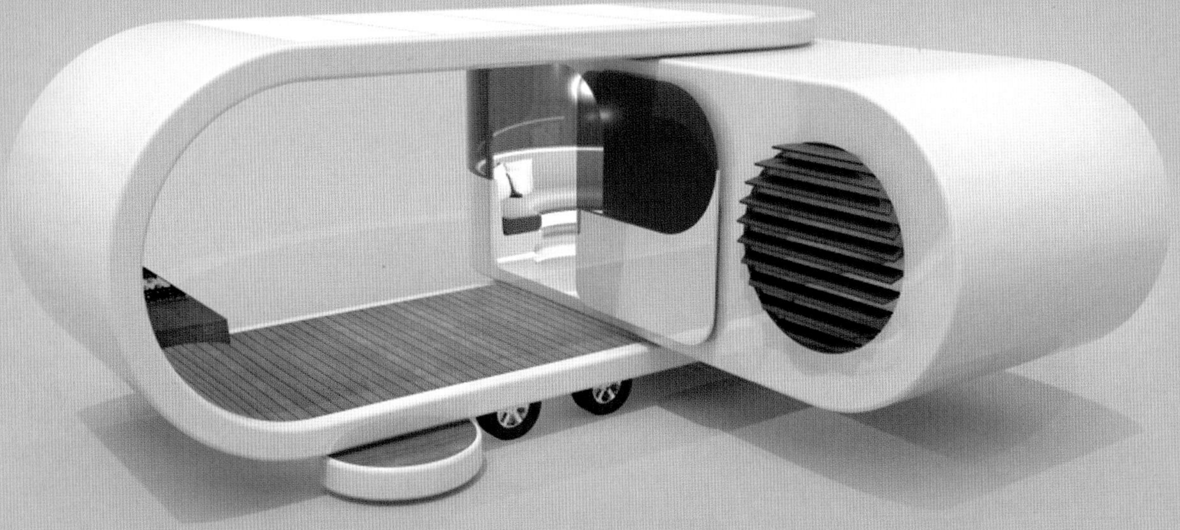

Above: The Romotow in its fully engaged state.

Looking like a giant USB memory stick being towed down the road, the Romotow is a new concept in caravans. Invented by architect and designer Matt Wilkie and engineer Stuart Winterbourn of Christchurch, it's lozenge-shaped and has an aerodynamically curved front, making it easier for towing. But when you park up, the real innovation comes into play. The whole thing swivels open, forming two wings arranged in an 'L' shape, one opening into the other. The result is a stunning design giving almost twice the space of a standard caravan, with a bedroom that flows beautifully out to an enclosed dining/kitchen area.

When you're ready to drive away, collapse the kitchen and dining furniture and swivel the caravan closed like a pocketknife. The question is, 'When does an invention become an invention?' Because as yet there is a website with an avid following of believers, there are designs for four different styles, three staff, an article in *Wired* magazine and even a US patent but, at the time of writing, no product or even prototype. Romotow says they're coming, though, so perhaps by the time you read this, the Romotow will be ready for the road.

The whole thing swivels open, forming two wings arranged in an 'L' shape, one opening into the other.

The Cadac Engine

A NEW KIND OF ELECTRIC MOTOR

In the first edition of this book, we began the story of a new electric motor like this:

> Wellington Drive Technologies are not based in Wellington, but on Auckland's North Shore. They hold a huge amount of intellectual property based around their new kind of electric motors, and they are poised to cash in big time. All going well, according to the CEO Ross Green, the company could nab 1 per cent of the world's US$15 billion market by 2003 – possibly sooner. This sounds like an amazing success story, but in fact the company has so far failed to cash in on what is a brilliant technological lead on the rest of the world. It is a lead they have had since 1987, but to date the company has made no money.

Updating the book gave us a chance to revisit the story of Wellington Drive Technologies (still based in Auckland). But first, here's a bit more background on their invention.

The basic principles of the AC electric motor – like that of the car engine – have not changed since Nikola Tesla invented it in the 1880s. Warkworth engineer Peter Clark turned the structure of the electric motor on its head. His Cadac motor was lighter, with less iron content, and was more efficient. In short it was revolutionary. Sinking money into development, he tested his invention successfully but couldn't find a market for it.

Small motors make up the largest part of the motor market, so Wellington Drive Technologies (WDT) focused their development in that direction. By 2000 the company could boast motors that were up to 30 per cent cheaper to produce and more efficient than the competition. Furthermore they were largely made of plastic, so were shapeable, allowing the motor to be built into an appliance rather than the appliance being built around the motor. Adding to this technical firepower, WDT made breakthroughs in the electronic control of the motors and turned their attention to the food processor, air-conditioner and vacuum cleaner markets.

At that time the WDT CEO told us, 'an unsuccessful product is 99 per cent right'. They continued to make no money.

When we contacted WDT in 2014 to ask how it was all going, Chief Technical Officer David Howell said, 'We're still in business!' When did the company begin to turn a profit? That finally happened in June 2013. Their secret was further narrowing their focus. In 2007 they dumped all the lines that weren't earning, largely gave up trying to license their technology to overseas companies and concentrated on what Howell calls the 'one technology that was giving us all the profit and none of the grief'.

After 28 years of development, the new motors little resemble the breakthrough technology that seemed so exciting in 1985. Most of the early patents have simply been left to lapse. But like Clark's first invention, the new motors contain far less iron and are far more efficient than their competition. And WDT's differential advantage is still the electronic control they were so excited about in 2000.

Wellington Drive Technologies employ 70 people in Auckland and outsource the manufacturing to Asia. They are now the world leader in the electric fans that blow air around large vegetable refrigerators in supermarkets and those big fridges that hold soft drinks in dairies. It might not be 1 per cent of $15 billion, but at about $25 each they have sold a million of them and now have 55 per cent of the world market in that niche.

With 28 years between inception and profit, 'We are the slowest burning start-up in the history of New Zealand,' says Howell. Next time you grab a nice cold L&P from the dairy fridge have a listen for the fan and spare a thought for the 28 years it took to get there.

The Barmac Rock Crusher

BREAKING ROCKS NOT SO BAD AFTER ALL

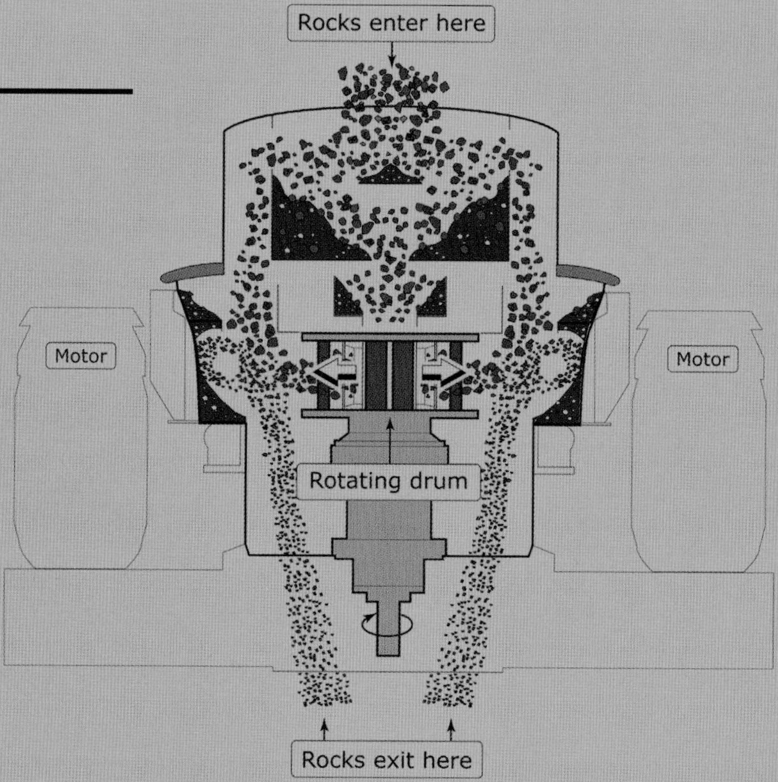

1970

In 1970 MacDonald began work inventing a new rock-crushing machine for the quarry in the Ngauranga Gorge.

Right: The Barmac rock crusher in action, crushing.

It used to be that breaking rocks was seen as a punishment for criminals, but it's been anything but for two New Zealand inventors.

George James (Jim) MacDonald (b. 1921, d. 1982) had a distinguished career in the Royal New Zealand Navy during and after World War II, winning the DSO and DSC, both with bars, and inventing a number of pieces of useful naval equipment. Among his inventions for the navy were a new type of slide rule and a device for firing torpedoes abreast, rather than one after the other. He was a senior officer from a young age and was described as a 'likeable and fearless young New Zealander…Cool and level-headed, he has the knack of making quick decisions and

2

The machine they came up with was born of the idea that the most efficient way to crush rocks was not to bang them against steel, or grip them in a vice – the two prevailing techniques at the time.

Left: The gravel road to Cape Palliser, Wairarapa.

inspiring his flotilla mates with a fine confidence… His leadership, cool courage…are an inspiration.'

After the war it must have been difficult settling back into civilian life, but MacDonald turned his inventiveness to more public works – literally. Studying hard to get himself trained, MacDonald became an engineer for the Wellington City Council, and in 1970 began work inventing a new rock-crushing machine for the quarry in the Ngauranga Gorge.

Bryan Bartley was also an engineer who worked for Winstone Aggregates, and when he came across MacDonald and his nascent machine he recognised a great idea that needed further refining ('scuse the pun), but the potential to be very big. Bartley brought some key ideas on how to improve the machine and thus he and MacDonald began a business partnership and friendship that would last many years.

The machine they came up with was born of the idea that the most efficient way to crush rocks was not to bang them against steel, or grip them in a vice – the two prevailing techniques at the time. Each of these ways had the major problem of wear and tear: over quite a short amount of time the action of the rocks against the steel would wear the parts away. MacDonald and Bartley's machine used the action of rocks on rocks.

Basically the machine they devised, dubbed the 'Barmac crusher' – see if you can work out why – relied on two principles: stones will break if you bang them together hard enough, and the steel parts will be protected from the abrasion if they are covered with a layer of rocks. The parts of the machine that did have to grate against the rocks were replaced with tungsten carbide, and the Barmac crusher became one of the most efficient and popular crushers on the market.

Rock crushing, believe it or not, is big business. MacDonald and Bartley saw their machine being used in the creation of aggregate (small rocks) for roading, but it also became a huge deal in the mining industry, where today the Barmac crusher remains the crusher of choice. Just one large machine is able to crush over 1 million tonnes of rock per year. It took eight years to turn their idea into a viable business, but the success of the product soon saw Bartley and MacDonald licensing its manufacture around the world, before finally selling the business in 1994 to a large international company who were delighted to get the Barmac into their product catalogue. Despite the number of copies and rip-offs, the name Barmac VSI (Vertical Shaft Impactor) lives on.

Unfortunately MacDonald died in 1982, but it seems Bartley can't shake the inventive spirit. After selling Barmac he went on to help develop another invention, the Kiwi Feather Prop, a self-feathering propeller for boaties.

The Britten Motorcycle

RADICAL MACHINE FROM NOWHERE

Other countries are good at and renowned for quite refined and artistic things – like fine coffee-table design, and the world's most elegant china or lace bedspread manufacturers. In New Zealand we reject this. We like motorbikes.

In 1994 the name of John Britten (b. 1950, d. 1995) joined those of Richard Pearse and Bill Hamilton (whom Britten named as his inspiration) as New Zealand's most famous inventors. Sometimes fame outstrips deserving, but in this case his exploits are, if anything, more incredible than the legend surrounding him. Throwing away the motorcycle manufacture manual and starting completely from scratch in terms of design and materials technology,

4

Starting completely from scratch in terms of design and materials technology, John Britten and his small team built the fastest four-stroke superbike in the world.

Left: The world's most advanced motorcycle, by John Britten

John Britten and his small team built the fastest four-stroke superbike in the world. On the way they pioneered technology that is now being exploited by motorcycle manufacturers everywhere.

At a motorcycle race meeting at Ruapuna Speedway in Christchurch in the early 1980s, John Britten and his friends were told they weren't allowed to race their motorcycles in the races for Japanese bikes. They were told to 'F off and start your own club', so they all rode off to Governors Bay and did just that. That was the start of the British, European and American Racing Series, or BEARS. A new category was born, and guys started building V-twin 750cc motorbikes to race in it. Japanese motorcycles weren't allowed.

Big race clubs all over the world picked up on the BEARS idea and it became a world phenomenon. By 1995 there was a world series with 11 rounds in Europe and America. John Britten had invented the format, and then he went on to invent the motorcycle to dominate it. The Britten bike soon smoked all other bikes of its class.

What Britten and his team of engineers came up with in the garage was a revolutionary motorbike, and its racing results were unprecedented. For five years in a row a Britten won the Sound of Thunder (previously BEARS) at Daytona. Britten's bikes have set four world speed records for bikes 1000cc and under in ordinary race trim, and held the outright speed record for Daytona. The world flying mile record (302.82km/h), the world standing quarter-mile (10.759sec, 134.617km/h), the world standing kilometre (19.33sec, 186.245kmh), and the world standing mile (27.135sec, 213.512km/h) were all set in New Zealand by Britten bikes. Britten achieved a long-standing ambition, winning the New Zealand superbike road racing championship series with a bike designed and made in New Zealand.

And the praise for Britten's achievement around the world is even louder than we New Zealanders heard: 'It's the world's most advanced motorcycle, and it's not from Japan, Germany, Italy or America,' shouted the cover of American motorcycle magazine *Cycle World*. The glossy cover screamed 'Stunner! – Britten V1100'.

Quite aside from race results and praise, the ultimate compliment is being paid to Britten in the design rooms of bike makers around the world. Britten's design has been copied and adapted to production motorbike manufacture. Triumph and Bimota raced each other to produce the world's first production bike with a carbon-fibre swing arm. Triumph copied the ducted cooling system. Computer engine-management systems are now becoming standard on V-twins, and axles around the world mimic the tubular construction of the V1000.

Britten's goal was never to manufacture a lot of products – he built just 10 motorcycles, all now owned by collectors or museums. For Britten, manufacturing took the creativity out of design – demanding a tight, disciplined approach to produce an item for a marketable price. Britten said, 'New Zealand is too far away from most major markets to succeed in mass production. But I strongly believe there is a niche market for Kiwis to exploit at the high-quality, low-volume end of manufacturing. It has to be top shelf.'

John Britten died in 1995, but enthusiasts and fans of his work keep his memory alive, and one of his V1000 bikes adorns the main hall of Te Papa in Wellington. Years after he completed it, the Britten bike was still the fastest motorcycle in the world, and for many fans of motorcycling the Britten V1000 will forever be the greatest motorcycle ever built anywhere in the world.

John Britten had invented the format, and then he went on to invent the motorcycle to dominate it. The Britten bike soon smoked all other bikes of its class.

The Amphibious Car

IT GOES ON THE SEA AND THE LAND

Terry Roycroft has called his invention the Sealander, because it goes on sea and it goes on land. While that does not seem a very inventive name, imagine an affordable, mass-produced amphibious car that could drive at motorway speeds on land, then drive straight into the water where it can ride up 'on the plane' at jetboat speeds. It only has to be half as popular as those ridiculous jetski things to be a huge success. The reason the world has never seen such a thing is not because people wouldn't want one, but because to make a viable commercial amphibian like this you must solve some very difficult engineering design problems. If you have ever sat behind someone towing a boat to Lake Taupo at about 50km/h on the Desert Road then like me you'll be saying, 'Please, please be a success!'

Like many successful inventors, Roycroft was, at first, just trying to solve a problem for himself, and make something to impress people with. Living on a farm on the Awhitu Peninsula south of Auckland, the engineering contractor could see the city just a half-hour boat ride away, but to drive there took an hour and a half. New Zealand's remoteness from the world is often credited with spurring invention, but in this case it was Waiuku's remoteness from Auckland. In the early 1980s, with time on his hands for a major project, Roycroft decided against agricultural ideas and chose the Sealander. He let the idea brew in his head and in little sketches for almost 10 years until, in 1991, he was ready to turn it into aluminium reality.

Like many successful inventors, Roycroft was, at first, just trying to solve a problem for himself, and make something to impress people with.

There have been many amphibious vehicles, starting with wartime landing craft, and including a very good Volkswagen model based on the Beetle, of which thousands were produced during the war to help the Germans try and fight the Russians. But none of these amphibious vehicles has ever been able to rise up to the plane in the water. What was lacking was a wheel retraction system that not only got the wheels out of the way to allow the craft to go more than about 5 knots, but could also be cheaply mass-produced. When Roycroft solved that problem, he knew he had something 'pretty damn good'.

The system Roycroft settled on was based on the normal wishbone suspension that all cars have. It was a cheap and simple solution that disengaged the axle and lifted the wheels above the waterline. The car part of the Sealander had mostly Subaru running gear, which Roycroft chose for its good 4WD setup. The car had to be much lighter than a normal car, or it would be a pig in the water, so Roycroft got the Sealander down to less than half the weight of a normal sedan by use of an all-aluminium monocoque

1993

The first prototype, completed in mid-1993, was entirely made in Roycroft's home workshop at Awhitu.

Below: Roycroft's Sealander

(no chassis or spaceframe) design. The normal car engine – initially a Subaru 1600 but later a 2 litre – was disengaged from the wheels and connected to a HamiltonJet (see page 90) drive unit when the wheels were raised. The steering wheel steered the wheels on land and the jet unit when the craft was in the water, where it could achieve a very fast 30 knots. The first prototype, completed in mid-1993, was entirely made in Roycroft's home workshop at Awhitu.

Now comes the part of the story that is repeated, in different versions, throughout this book. Roycroft says, 'In my naivety I thought if I could get it going, there would be no trouble attracting capital because it was so unique and exciting.' He filed the patent himself – 'a whole story on its own' – and made some publicity material, attracting attention from New Zealand television and print media. That was the start of what Roycroft admits were 'three or four years that were not a good time'.

Finally Roycroft hooked up with investor Erny Yeoman and began in earnest to get his working prototype to commercial level. Meanwhile, he learnt a lot about the craft by driving it, and even more by letting his sons drive it. They almost sank it several times including once with a camera crew on board. Roycroft kept an empty 4-litre container attached to a long rope on board in case the Sealander ever went down so that they could locate it for salvage – the first prototype had no airtight compartments and would 'sink like a rock'. Roycroft's sons have got the knack of hardly even slowing the amphibian down as they approach the water, just timing the wheel retraction perfectly so that it hits the water at 50km/h and jets right up on the plane. The Roycrofts especially like to pull this move right next to someone struggling to back their trailer down the beach to launch their motor boat.

Little did Roycroft know at the time this was just the start of the Sealander's story. For the next chapter, read on…

Gibbs Aquada and Quadski

PUTTING THE SEALANDER INTO TOP GEAR

23

The honour of being the world's first commercialised amphibious vehicle that can be driven at speed on land or sea, 23 years after the Sealander first pulled up its wheels, goes to the Gibbs Quadski.

Above: The Quadski.
Far right: The Aquada. It's well known that the red ones go faster.

In 1995 Alan Gibbs was an extremely successful businessman with an inventive instinct and a love of all things mechanical. He'd built himself an amphibious catamaran called *Ikarere* to cross the mudflats of the Kaipara Harbour north of Auckland, but it was too big and heavy. It wouldn't get up on the plane and its wheels got stuck in the mud and swamped the whole boat when the tide rose. Gibbs was fascinated by the idea of building the first commercially available amphibious vehicle, but his catamaran wasn't it. He abandoned his approach to 'amphibuity' when he came upon a photo of Roycroft's Sealander displayed in a Wellington art gallery in mid-1995. 'I realised Terry had made nice progress in his wheel-raising system.' Gibbs bought the patents from Roycroft in 1996 and immediately started working with him to develop the Sealander. 'At the start I asked myself, "Do I just do this and make toys for me, or make it a bit harder for myself and try to make it commercial?" I decided to do the latter.'

It wasn't easy. For the next 16 years Gibbs drove his teams constantly forward. It proved to be a massive task to commercialise the Sealander. Gibbs set up headquarters in the UK where his team created their own water-jet engine, invented a way to incorporate the suspension with the wheel-retraction mechanism, and designed a body that functioned perfectly and looked great. They also solved the all-important power-to-weight problem – which Gibbs says was the 'constant bugbear'. By 2012 they had spent $200 million on research, developed the original craft into nine completely different sorts of amphibians, and held over 100 separate patents,

It's a quad bike that goes on the water, and it's a jetski that goes on land.

but had not yet delivered a product to a customer. Gibbs was happy, though. 'I called it "golf" – my retired mates were all off playing golf, while I was off doing this, for the fun of it.'

A major milestone was the 2003 London launch of the Aquada to a curious world. A sporty three-seater, it was the sexy, grown-up version of the Sealander. Though he had only worked on the project for the first five years, Terry Roycroft watched from half a world away and reported himself 'chuffed'. It could do 160km/h on the road then drive straight into the water and do 50km/h. *Time* magazine named it among the best inventions of 2003. On 14 June 2004 it looked the part as Richard Branson swashbuckled across the English Channel in it – smashing the record for a crossing by an amphibious vehicle. But still the pieces wouldn't fall into place for a commercial release. United States safety regulators refused to certify it for the road because it didn't have airbags. Gibbs pointed out that airbags would deploy every time the Aquada hit a wave, but the Americans wouldn't budge. In the UK they hit another speed bump, or rough chop, when Rover decided not to make any more of the V6 engines used in the Aquada. Then, just before they were to deliver the first Aquada, Gibbs let the prospective owner borrow the boat for a weekend. Assuring Gibbs he was an experienced boatie, he promptly drove into the Atlantic Ocean off Land's End in the south of England, got into trouble and drifted halfway to France before being rescued by the coastguard. The Aquada was so good it allowed people to get into trouble. After building 45 sparkling Aquadas, Gibbs shut down production.

Gibbs then changed direction and developed other vehicles based on the same technology. The Phibian and the Humdinga are amphibious trucks with 300- and 500-horsepower engines respectively. They are designed for the 'first responder' market – search and rescue, fire, coastguard and surf rescue. Their prototypes have stirred great interest. But the honour of being the world's first commercialised amphibious vehicle that can be driven at speed on land or sea, 23 years after the Sealander first pulled up its wheels, goes to the Gibbs Quadski.

It's a quad bike that goes on the water, and it's a jetski that goes on land. Finally – a solution to the age-old problem of how to annoy people on sea and on shore! It's the first commercially available amphibian to do more than 10 knots on the water and more than 10km/h on land. In fact, with its 1300cc BMW engine it can do 72km/h on land and the same on the water, taking only five or so seconds to transition between the two.

The Quadski hit the market in April 2013 and since then the 175 staff at Gibbs' Detroit plant have turned out 100 a month. For just $42,000 you can buy a piece of New Zealand invention history. In February 2014 the Quadski was featured on *Top Gear* (the most watched factual programme in the world). In a race from one side of Italy's Lake Como to the other ('the usual bullshit', as Gibbs calls it), Richard Hammond drove Alpha Romeo's brand new 4C and Jeremy Clarkson drove Alan Gibbs' Quadski. Of course, the Quadski won.

Gibbs reckons you need to be a masochist to be an inventor because people are wisely cautious of anything really new. 'But it's a hell of a lot of fun,' he says, 'and it's kept me entertained for twenty years!' And what about Roycroft, creator of the original idea? 'Gibbs treated me very honourably,' he says. 'I'm very comfortable.'

6.

It Seemed Like a Good Idea at the Time

JOGGING, THE CHICKEN SWITCH
AND THE SPLITTING OF THE ATOM

Sometimes a nation has to put up its hand and just admit 'it was me'. Some things you might say the world would have been better off without but someone invented them anyway. These are those things. New Zealanders have invented a lot of stuff, and not all of it has been what you'd call top shelf.

It is interesting to ask what makes an idea 'good'. Some invention theorists like to pose a kind of evolutionary model, which is based on the Darwinian notion of the survival of the fittest. By 'fittest' Darwin meant the creatures that best fit in to (were best adapted to) their environment. Thinking about inventions in this way we can imagine that ideas or inventions that best suit their environment will survive. An idea might be right in one century or country, but later or elsewhere be wrong. This is why inventions come and go, why one thing replaces another; an invention that seems pointless now might have fit its time perfectly. This is why New Zealanders are so good at making inventions to suit New Zealanders – because only we know our environment. And this is also why an invention that can be used all over the world is more rare and special – because the environments it must be fit for are much more varied.

Some of the 'bad ideas' in this chapter, then, are inventions that seem wrong now but we can see how they were once perfectly fit for their time or place. Still, some of them just don't really seem like they could possibly have been a good idea anywhere, ever.

Rutherford and the Split Atom

BREAKING UP IS HARD TO DO

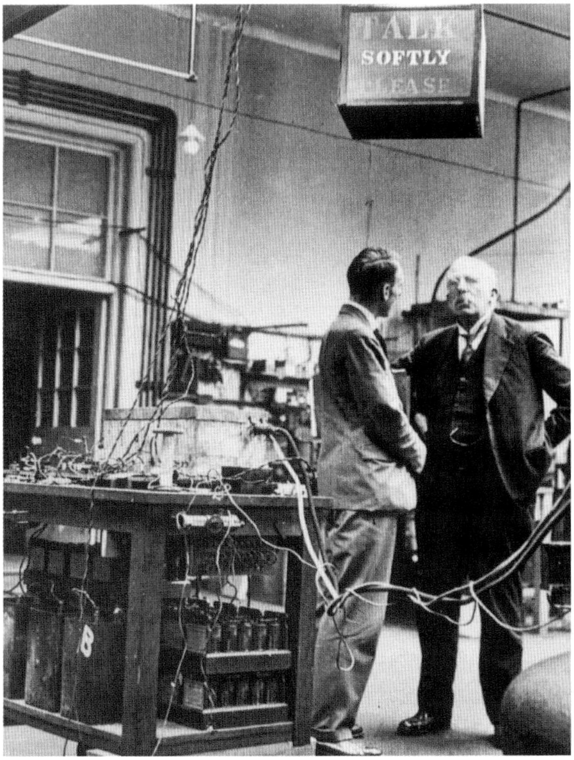

3

This experiment was Rutherford's third great achievement, and of course seeded the idea of nuclear fission, which others would go on to develop into a great boon for mankind, or a great threat to our survival, depending on how you look on it.

Right: Rutherford (right) was renowned for having a loud voice and special signs were installed in the Cavendish Laboratory to remind him to keep his volume low, so as to not disturb the delicate experiments.

In about 400BC, the Greeks conjectured that the world was made up of tiny particles, invisible to the eye. Different combinations of these particles gave rise to the vastly different materials that we see around us in the world. They called these particles 'atoms' – the Greek word for 'indivisible'. While other variations of this idea of the universe came and went in the following 2000 years, that basic view stuck until about 1850, when scientists discovered that the atom could come with differing electrical charges (positive or negative), and that these charges could change. It was then suggested that the atom must contain other smaller particles, but no one had a firm idea of how this all hung together.

Meanwhile, at least for the first 2000-odd years of this, Kiwi Ernest Rutherford wasn't around so couldn't set them straight. But then he was born in Nelson and the world would never look the same again.

Rutherford (b. 1871, d. 1937) was by all accounts a very clever child. He breezed through school, where he seems to have had the quintessential Kiwi education – head boy and in the first XV. He went on to Canterbury University, completing an under-graduate degree and an honours year and publishing a couple of scientific papers, which got him noticed overseas as an outstanding innovator in the forefront of electricity research. He won a scholarship and moved to England in 1895 and began as a research student at the distinguished Cavendish College in Cambridge – the first ever 'foreigner' to be honoured in such a way, and possibly the first example of a Kiwi having to leave New Zealand to continue his career. The beginning of the brain drain?

Rutherford did extremely well in this role, contributing a lot to the work at the laboratory, and gaining an invitation to work at McGill University in Montreal, Canada – a role he took up eagerly. It was at McGill that he made his first major discovery in science – that atoms can spontaneously transmute into other elements through radioactivity. It was this work in explaining radioactive decay that won him the Nobel Prize in 1908. Ironically the prize was for chemistry, not physics, yet Rutherford described himself as a physicist first and foremost. As a by-product of this research into radioactivity, Rutherford invented radiometric dating, suggesting it could be used to finally give an accurate age for the Earth – which it eventually did. He also invented the terms 'alpha ray', 'gamma ray' and 'half-life' to help describe radioactivity.

Returning to England in 1907 he worked first at the University of Manchester, then back at the Cavendish laboratory, where he made his next significant breakthrough.

Rutherford devised a plan to derive the internal structure of the atom. It went a bit like this (and by all means, do try this at home). Get a piece of gold foil. Beat it so it is as thin as you can make it – just a few atoms thick. Now grab your alpha particle generator and point it at the gold foil. If you've got this far, you'll know already – as was the prevailing opinion at the time – that an alpha particle is a large, heavy collection of stuff and so it should have no problem punching though a little old sheet of gold foil.

Start firing the alpha particles. Expected result: the particles should smash straight through the foil to the other side. Actual result: they do... mostly. But every now and then one bounces back. Now this was completely unexpected for Rutherford, who likened it to 'firing a cannonball at a sheet of tissue paper and having it bounce back at you'.

Rutherford deduced from this result that the atom, rather than being a single indivisible particle, was mostly just empty space, but with something very small and very dense in the middle – dense enough to send that alpha particle bouncing back. Most of the alpha particles pass right through the empty space, but every now and then one of them smacks into the bit in the middle – the nucleus – and is deflected right back. Rutherford had shown that the atom was made up of smaller 'things' arranged a bit like a solar system – small electrons orbiting a small nucleus. He had, theoretically at least, split the atom. It is hard to think of any discovery in all of science which was more important in the quest to find out what the universe is.

It is hard to think of any discovery in all of science which was more important in the quest to find out what the universe is.

Not content to finish there, Rutherford went on to *actually* split the atom – he bombarded nitrogen atoms with alpha particles and witnessed them split into oxygen. It was the first time a person had ever performed the alchemy of changing one element into another. This experiment was Rutherford's third great achievement, and of course seeded the idea of nuclear fission, which others would go on to develop into a great boon for mankind, or a great threat to our survival, depending on how you look on it. For Rutherford, it was just good science.

Rutherford's place as one of the greatest scientists of all time cannot be overestimated. Biographers write that he is to the atom what Darwin is to evolution, Newton to mathematics and Einstein to relativity. Rutherford's greatest achievement, however, may have been as a teacher and mentor to many of the next generation of eminent scientists – names like Chadwick, Bohr, Oppenheimer and Geiger all learnt their craft from the humble Kiwi with the big voice. Rutherford valued the simple explanation of his craft, saying at one point that 'an alleged scientific discovery has no merit unless it can be explained to a barmaid'.

Rutherford in his lifetime was awarded the Nobel Prize, 21 honorary degrees and countless awards, and was also named 'Baron of Nelson'. The story goes that as now Lord Rutherford lay on his deathbed in 1937, at age 66, he called for his wife to make a donation of 100 pounds to Nelson College, his alma mater. Today Rutherford's wish is honoured with his face adorning our $100 note.

Jogging

QUEEN OF THE POINTLESS SPORTS

Before jogging became popular, you just didn't see people out on the streets running around. Running was for getting somewhere when you wanted to get there a bit faster than walking but you didn't have a bike. People reasoned that if you ran out of a house, but then just ran back into the same house half an hour later, you hadn't gotten anywhere and there was no point. There was no such thing, then, as a running shoe, and even competitive runners would train for maybe half an hour a week. To you and me, this might seem like an ideal situation, this global lack of running around in humorous shorts. But a New Zealander saw fit to change it.

Arthur Lydiard (b. 1917, d. 2004) was an Auckland shoemaker, and in 1945 he decided to design a

1960

In one hour at the 1960 Olympics, two of Arthur's local athletes, Peter Snell and Murray Halberg, won gold medals, then another of his athletes, Barry Magee, got bronze in the marathon two days later.

Above: Coach Arthur Lydiard (centre), 56, takes part in a keep-fit run with members of the Wellington Joggers' Club in June 1974.

personal programme to keep himself fit. Nine years later he was the New Zealand marathon champion. He began to coach young runners using the methods he developed for himself. His basic credo was a combination of aerobic and anaerobic running, as often as two or three times a day for weeks on end, and his training regimes were well worked out over years of trial and error. The results speak for themselves. In one hour at the 1960 Olympics, two of Arthur's local athletes, Peter Snell and Murray Halberg, won gold medals, then another of his athletes, Barry Magee, got bronze in the marathon two days later. They were just the beginning. Lydiard's programme produced a string of Olympic medals and world records. More than that, he became a coach of coaches, the world's genuine guru of fitness. It is safe to say that today, no corner of athletics (or fitness training for any sport) is untouched by his influence.

But it's not that influence that concerns us here. When Lydiard came back from the 1960 Olympics as a medal-winning coach, he was asked by his brother-in-law to speak about his training methods at the Tamaki Lions Club. After the talk he was approached by some tubby businessmen who had had coronaries. The wisdom of their GPs said that they should no longer exert themselves. Lydiard said if they wanted to improve their condition they should embark on a programme of training. Then he took them out onto the streets and showed them how. Lydiard began speaking to people all around the country about his Olympic success and how it could be applied to everyday fitness. One day in 1962, Lydiard was flying back to Auckland from Christchurch with former Auckland mayor Colin Kay. Lydiard told him about his programme. Kay got together a group of about 20 businessmen, few of whom could run 100m at a stretch. They formed the Auckland Joggers' Club – the world's first. Eight months later eight of them ran marathons.

Ironically, at the time, Lydiard's runners were sponsored by the Rothmans cigarette company. When Arthur was speaking around the country he was working for the Rothmans, so he couldn't tell people to stop smoking. 'If I told them to stop smoking and drinking they wouldn't listen to me. I told them to exercise and most of them stopped smoking and drinking a few months later.' He used people's natural competitive instincts in non-competitive training, telling businessmen instead of running a circuit around the neighbourhood to try

People reasoned that if you ran out of a house, but then just ran back into the same house half an hour later, you hadn't gotten anywhere and there was no point.

running in a straight line and getting their wives to come out in the car and pick them up after an allocated time. The men would run as hard as they could in order to look good.

In the summer of 1963 American track coach Bill Bowerman visited New Zealand with his athletes, as the guest of Lydiard. As well as sharing training techniques, Lydiard one day took Bowerman jogging. The word was new to the American – a 'jog' was what Lydiard called the run you did on an easy day's training. Bowerman found himself lacking in fitness and when he got home to Oregon he set up a local jogging class to get himself in better condition. Bowerman was helping to build a vast and enduring market for his own product, because he was the originator of the waffle-soled sports shoe. (Arthur Lydiard also said, 'I taught Bowerman how to make shoes.') The running craze took off and spread around America and the planet. The shoe company became Nike, the biggest sports manufacturer in the world. Bowerman wrote the best-selling book *Jogging* and received a special medal from President Kennedy. His comment: 'I am but the disciple. Arthur Lydiard of New Zealand is the prophet.'

Today Auckland remains the centre of world jogging. The Round the Bays run is one of the world's largest annual fun runs with some 40,000 participants each year. Lydiard's most basic message was that cardiovascular exercise is good for anyone, no matter what kind of body they have. 'People's bodies are basically all the same. Anaerobic and aerobic exercise affects all people the same way. It doesn't matter how fast or slow they are, the running can give them endurance.' Lydiard never gave up on spreading the fitness message: right up until he died he was still travelling the world to talk about training for fitness. Lydiard – runner, coach, inspirational speaker, author, recipient of New Zealand's highest honour (Order of New Zealand) – died in Austin, Texas in 2004 while on a lecture tour at the impressive age of 87, of a suspected heart attack.

The Telegraphic Typewriter

WHY WE ARE STILL SEEING QWERTY KEYBOARDS TODAY

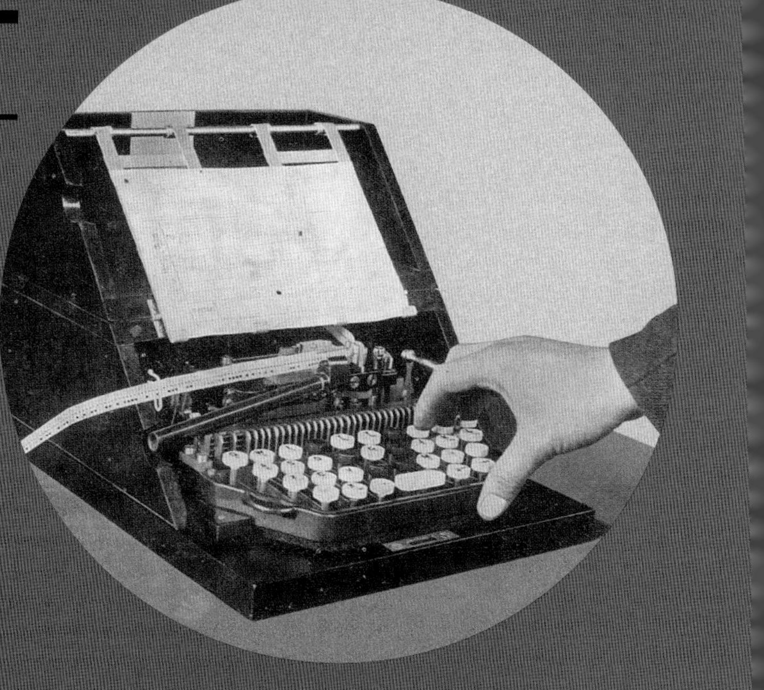

1899

Those great iconic images of a smoky newsroom full of stressed-looking journos pacing frantically, and waiting for the 'clack clack clack' of the telegraphic typewriter in the corner, are thanks to the invention he patented in 1899.

Right: The typewriter in action with the familiar keyboard layout.

If you are over a certain age, you will know what a typewriter is. If you are under a certain age, imagine a computer, with a keyboard, a screen and a printer. Now throw away the screen, and make it that whatever you type on the keyboard just comes out on the printer. That's a typewriter. Yes, it is as naff as it sounds, but that's what we all used to use before computers came along, so show some respect.

The thing about typewriters in the late 1890s is that they were all so *local*. The person typing sat in right front of the typewriter and the paper came out of it right there. If they were – for example – a journalist, as was Donald Murray (b. 1865, d. 1945) then that journalist would have to frantically type their story at the keyboard, rip the finished story from the back of the typewriter (with that familiar sound) and run down to the editor's office yelling, 'Hold the presses!' Not ideal, but at least that worked if everyone was in the same place. When Murray was working at the *Sydney Morning Herald*, receiving and filing stories from all over Australia and the world, it got even worse. He'd have to take his finished story to a telegraphist, who would convert it to Morse code, then laboriously tap it out, so someone at the other end would listen, note it down in Morse, and convert it to English, and retype it. Horrifically inefficient.

Simply put, Murray's brainwave was to separate the printing bit of the typewriter so that it could go anywhere, meaning you could type at one end and the remote printer would print out what you'd written. In this way, he got rid of all the work and

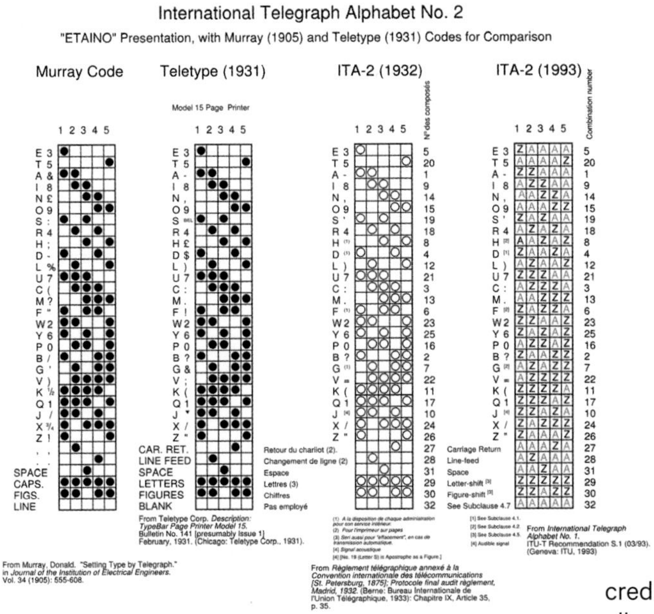

Left: The Murray Code was simpler to read than its predecessor.

While the keyboard is immediately familiar to us today, the machine used a particular code when it printed that Murray also invented – the Murray code.

delays in the middle, including those telegraphists, which is probably why they aren't showing up in my spellchecker anymore. You certainly don't meet a lot of telegraphists at barbecues these days, and Murray is to blame for that.

Murray was another son of Invercargill, who attended university at Canterbury College School of Agriculture (now Lincoln University) near Christchurch. His move into journalism happened in 1887 when he worked at the *New Zealand Herald*. His invention changed the nature of the newsroom. Those great iconic images of a smoky newsroom full of stressed-looking journos pacing frantically, and waiting for the 'clack clack clack' of the telegraphic typewriter in the corner are thanks to the invention he patented in 1899. It became extremely popular and was introduced into newsrooms and commercial applications worldwide. So popular, in fact, that we'll skip to the end for a second and tell you that Murray retired to Monte Carlo – the Invercargill of Europe – in 1925.

There is more to tell on the machine itself, though. Firstly – and students of alternative keyboard layouts will appreciate this – Murray's invention is credited with cementing in the QWERTY layout of all modern keyboards. At the time that Murray was developing his invention there were other keyboard layouts in use. Murray wanted to make his machine as easy as possible to use, so he chose a popular layout – QWERTY. Murray's telegraphic typewriter became so ubiquitous that the QWERTY keyboard it used graduated from being just popular to become the default standard, and it remains so today.

While the keyboard is immediately familiar to us today, the machine used a particular code when it printed which Murray also invented – the Murray code. This new way of encoding characters for transmission became the standard at the time – the International Telegraph Alphabet No2 (ITA2) which lasted until 1963 when it was superseded by something we may be more familiar with, the American Standard Code of Information Interchange – or ASCII. Murray code was adapted from a previous code – the Baudot code – but Murray enhanced and improved it. The code is still in use today in some applications, such as amateur radio. When you typed at the keyboard, a strip of paper would advance, and a series of five dots would print. The system was sophisticated enough to transmit characters and letters, and was quite readable once you got used to it.

In his later life, while sitting in quiet comfortable contemplation, Murray became quite a philosopher and wrote a number of books on the philosophy of power. Those books are largely forgotten, but his contribution to the world should not be. After all, with the transmission and remote printing of a message that his invention enabled, couldn't we say he also invented email?

Microclimate Farming

EARLY MĀORI FIND WAYS TO GROW KŪMARA IN THE COLD

Right: Kūmara pits on Mt Hobson, the site of an old Māori settlement.

When the migrating Polynesians got to New Zealand, they were struck by an immediate and deadly serious challenge: to survive they had to grow the foods they'd brought with them, or find new foods. Their disappointment in the fruits of New Zealand's forests is recorded in the name they gave our only native palm – the 'nikau', which means 'no coconut'.

Things would not be fun if they couldn't make the kūmara grow. But it is a vegetable that comes from a warmer land (tropical South America) and though it grew well in most of Te Ika-a-Maui, growing it in Te Wai Pounamu was a harder job. So the early Māori invented a new sort of agriculture. They started by building wooden fences and stone walls around the plantations. This kept the wind out, but they also realised that stone walls had another benefit – soaking up the warmth of the sun during the day, and radiating it back out in the evening, lengthening the period the crops were kept warm. Most of the stone walls built by early Māori were built for this reason, and you can still see them in places like the Otuataua Stonefields in Mangere, Auckland.

But the real innovation was to mix stones in with the soil. This way of effectively turning temperate soil tropical was a uniquely New Zealand invention.

But the real innovation, an invention anthropologists call 'lithic agronomy', was to mix stones in with the soil. This way of effectively turning temperate soil tropical was, according to many academics, a uniquely New Zealand invention.

Thousands of hectares of kūmara-growing land around the country have been investigated. Gravel, brought from elsewhere, has been mixed with the soil. The gravel particles are between 20mm and 60mm in diameter. In some cases the edges of the gravelled areas are perfectly straight lines, showing that the addition of stones was done on purpose by humans.

Lithic agronomy allowed the settlers to thrive and move further south to colonise more of the land we call Aotearoa.

Biospife

IS IT A KNIFE? IS IT A SPOON? YES

2

The real benefit of the spife is that it can be used as an example of a 'portmanteau', a combination of two words to create a new one.

Right: The biospife, massacring innocent kiwifruit.

Zespri – purveyors of fine kiwifruit since 1997 – created the 'spife' a few years ago. It's like the child you'd get if you mated a spoon and a knife, and continues the tradition of cutlery grafting made popular by the 'spork'. The real benefit of the spife is that it can be used as an example of a 'portmanteau', a combination of two words to create a new one. A good example of a portmanteau also describes the spife – blintriguing.

Not content with grafting part of a knife onto a spoon, the design gurus in the sunny Bay of Plenty have now achieved the culmination of their cutlery cross-breeding with the creation of a bioplastic spife, or a 'biospife'. It's fully compostable and made with bioplastic infused with kiwifruit material. Like its cousin, it's used to both cut and then scoop kiwifruit so that one may consume their inner deliciousness. Why the age-old technique of squeezing the outer skin while holding near your mouth is not adequate is unclear. However, having said all of that, my kids love the spife and were quite clear that they considered it much more important than any of the other inventions in the book, so here it is.

I wonder what mixture of eating utensils will come next? The 'foon'? The 'knork'? Heaven forbid, the 'chopstork'?

Zespri – purveyors of fine kiwifruit since 1996 – created the 'spife' a few years ago. It's the bastard child of a spoon and a knife, and offensive to both parents.

Stamp-vending Machine

OUR FIRST TECHNOLOGICAL BREAKTHROUGH

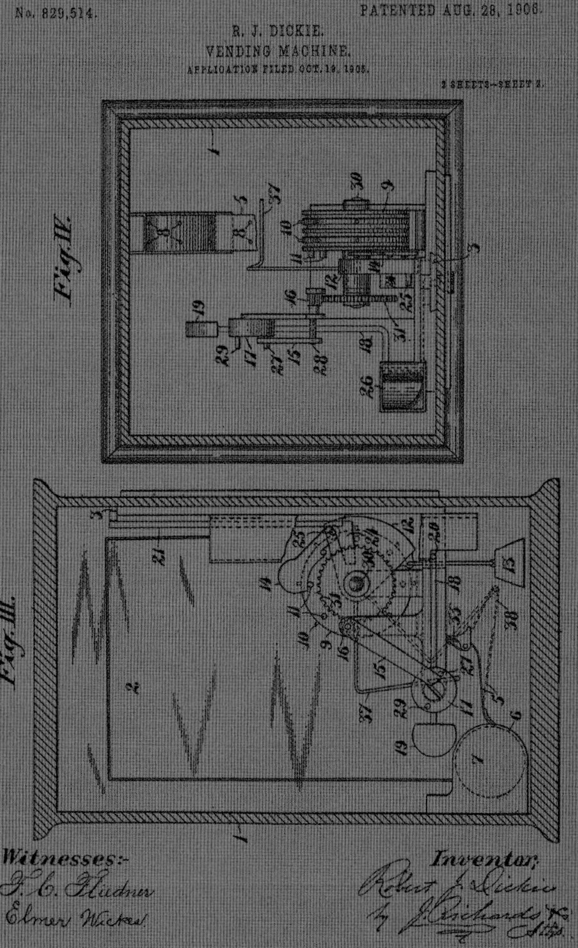

3902

The first public trial was an immediate success, and 3902 stamps were sold in 14 days.

Far right: R. J. Dickie (left), his partner J. H. Brown (standing), and W. Andrews (right), the engineer who produced the first working model, photographed with the first Dickie stamp-vending machine, 1905.

In AD60 Hero of Alexandria invented a water-vending machine to sell water to thirsty people. In 1840 the postage stamp was invented in Britain. It was only a matter of time until these two inventions, racing towards each other through history, collided to form the stamp-vending machine. When they did, it was a hero of New Zealand who brought them together. It occurred to 15-year-old Robert James Dickie (b. 1876, d. 1958) while working at the front desk in a Wellington post office in 1891 that a machine should be able to tear stamps off and hand them to people and save him from doing it. When Dickie saw his first moving picture images he reasoned that stamps could be handled just like photographs strung together. He began to plan, and over the next several years the idea gestated in his head.

At age 28, Dickie – who was hopeless at drawing – roped in a couple of helpers, draughtsman JH Brown and engineer W Andrews, and together they designed, built and patented the first ever stamp-vending machine in the world. It wasn't very big – the mechanism was only 22 centimetres by 10 centimetres. A fluted sprocket wheel with weights attached was set in motion by the descending coin, making a single stamp project from a slot.

Several American stamp machines had been invented already, but Dickie's had a superior mechanism. The American machines were impractical because they were heavy, expensive and had a

Dickie once visited a factory in England where 800 men were making machines from his design.

very limited stamp-holding capacity. Dickie's major innovations were the fact that the falling coin caused the stamp to be ejected without pulling a lever or turning a wheel, and that the stamps were stored in a roll. The roll of stamps had been used before on machines to fix stamps automatically to multiple envelopes, but never in a vending machine. The Dickie patent was later licensed for machines dispensing tram and theatre tickets.

On 15 June 1905 the first public trial took place and a Wellingtonian (whose name remains unrecorded) was the first person to buy a stamp from a vending machine in the vestibule of the Chief Post Office. It was an immediate success, and 3902 stamps were sold in 14 days. Then another feat of Kiwi ingenuity brought the trial to an end. Someone figured out that you didn't need to use a penny, just something that was shaped like a penny, in order to score free stamps.

At that time it seems stamps were a hot area for innovation. In 1904 Ernest Moss of Christchurch invented a machine that might have put Dickie's out of business. Moss's franking machine, patented in 1905, meant an envelope could receive a 'frank mark' – a rubber stamp of a certain design to replace the stick-on one. For a number of years from 1905 onwards, New Zealand was the only country to use the franking machines (or more correctly postage meters) as an alternative to the stick-on stamps, and we still hold the record for the country with the longest continual use of postage meters. It's surprising this fact is not celebrated with a national holiday and taught to children in school.

However, luckily for Dickie, it seems there was room in the market for both ideas – particularly because Moss's franks were only valid for domestic mail – and Dickie's invention was an international success. A year after the first machine proved itself, Dickie travelled overseas, taking his machine with him on a business trip for the Post Office. In 1907, to get the British to cotton on to the idea, he set up a machine in the lobby of the House of Commons. It was a popular curiosity. A story from the *Dominion* at the time of Dickie's death says, 'Frock-coated members stopped and stared as they passed through the lobby… like children at the fair, the venerable Edwardians could not resist the temptation to try out the world's first practical stamp-selling machine.'

The marketing ploy worked and the British government began to manufacture the machines. Soon the New Zealand stamp-vending machine was being manufactured all over the world under licence from Dickie. In fact, by the time the New Zealand government decided it wanted some more in 1909, it could already import them from Britain. Meanwhile, at the 1909 Alaska-Yukon-Pacific Exposition in Seattle the vending machine won the Gold Medal, Grand Prize and Diploma against all comers. Interestingly, Dickie only just beat a Japanese man to the line. There still exists in Japan an automatic stamp- and postcard-vending machine built in 1904 – the year before Dickie's machine went into action. It was made by Koshichi Tawaraya and bought from him by the Japanese government. It never worked too well, it wasn't based on the same principle that Dickie succeeded with, and it never went into public operation.

Dickie kept updating his designs as time went on, and his stamp-vending machines were manufactured all over the world for 50 years. Just after World War I, the company that was manufacturing the machines had orders for over 100,000 of them – orders they couldn't fill. Dickie once visited a factory in England where 800 men were making machines from his design. By 1938, 18,000 had been installed in Britain alone! For half a century Dickie was the world king of stamp vending, although he never quit his job at the Post Office.

The High-speed Dental Drill

MAKING THE MURDER HOUSE MORE TOLERABLE

Contrary to what you might think, yanking an offending tooth from your jaw wasn't the only dental technique that primitive humans had. There is evidence that as long ago as 7000BC, dentists were using bow drills to grind the decay from teeth. It must have been pretty painful, as low-speed drilling (for those of us who had that inflicted on us) is a noisy, smelly and ultimately unpleasant experience. When I was a kid we called a visit to the dentist 'going to the murder house'. We believed dentists, and especially dental nurses, were pure evil.

However, it needn't have been that way. In 1949 New Zealand dentist John Patrick Walsh

400,000

Walsh's drills typically spin at 400,000 revolutions per minute (rpm), versus the previous low-speed drills, which operated at about 3000 rpm.

Above: Sir John Walsh, shown here at his desk at the Otago Dental School, was dean of the school for 25 years from 1946.

Above: The first air dental drill.

(b. 1911, d. 2003) worked with lab staff in Wellington and created a drilling handpiece that was driven by compressed air. The air is forced pneumatically through to the handpiece where it passes through a turbine to spin the handpiece at a tremendous rate. This new handpiece was so well designed that it has been universally adopted and remains virtually unchanged today. Walsh's drills typically spin at 400,000 revolutions per minute (rpm), versus the previous low-speed drills, which operated at about 3000 rpm. That extra speed translates to quicker drilling, and less patient discomfort.

Walsh – later knighted for his services to dentistry (I wonder if he got a plaque?) – was born in Australia but adopted New Zealand as his home, and made a number of contributions to Dunedin, where he was based and helped establish the University of Otago's School of Dentistry. He also changed the way dentists worked with their patients, seeing them as people who needed holistic treatment, not just a set of mandibles and canines.

Walsh's spirit of pioneering dentistry lives on. Kiwi dental company Triodent was founded in 2003 by dentist and inventor Dr Simon McDonald. Triodent grew from $250,000 in capital to achieve $90m in exports in 10 years, and builds on Walsh's tradition by manufacturing innovative dental products like the Wedgeguard – a small device for protecting the adjacent tooth while a dentist is drilling in a decayed one.

Walsh's real legacy is the current generation of New Zealand kids who have no idea what the 'murder house' is, and for whom a trip to the dentist is not laden with fear and pain, thanks in part, at least, to Sir John Walsh and his high-speed drill.

But Walsh's real legacy is the current generation of New Zealand kids who have no idea what the 'murder house' is, and for whom a trip to the dentist is not laden with fear and pain, thanks in part, at least, to Sir John Walsh and his high-speed drill.

The Thermette

TURNS THE WORLD INTO YOUR KITCHEN

What makes the Thermette brilliant is that it is perfect. In the 85 years since it was invented, no improvements have been made on it. It has been manufactured under the same patent continuously for that whole time and it remains, as it was, the quickest, most efficient way to boil water in the outdoors.

The inventor's name was John Ashley Hart (b. 1887, d. 1964) and he was originally, like many great things, from the Manawatu. Most of Hart's 32 other patents are now forgotten, but the Thermette, invented in 1929, caught on as standard equipment for New Zealand troops during World War II. The army approached Hart to ask if he would waive the patent to help in the war effort, and he agreed. The small round scorch marks it left on the earth at first confused the German troops all over North Africa, where the Thermette gained its army nickname, the 'Benghazi Boiler'. Soon everyone knew the scorch marks were a sure sign that Kiwis had been there.

The Thermette can boil enough water for 12 cups of tea in just five minutes, using any old rubbish as fuel. 'The stronger the wind, the better it boils,' was one of Hart's early slogans, because wind sucks air

1931

The original Thermette was first sold in 1931 in a blue, green and orange tin, or in tinned copper if you had a few extra bob.

Far left: A copper Thermette in action, making a cuppa for one of the authors (right) and a homeless man (left).

In the more than 80 years since it was invented, no improvements have been made on it.

up through the conical chimney inside the boiler from the base where the fire is lit. The sucking action makes the fire roar, and the heat is transferred not only to the base of the Thermette, but through the heated air rushing up the internal chimney. No heat is wasted, and that is why the Thermette is so efficient. Its efficiency makes it an environmentally sound product, and it uses no pollutant gas, petrochemicals or hydrocarbons.

The original Thermette was first sold in 1931 in a blue, green and orange tin, or in tinned copper if you had a few extra bob. MS Services Ltd in Auckland is still making them to the same design. It used to be that any council workers, postal workers, and telegraph men on the side of the road could be seen setting up their Thermette, but they have largely disappeared from our roadsides. At the height of their popularity tens of thousands were made a year; now it is just a few thousand, but Trevor Tull at MS Services is hopeful for a resurgence in popularity. They are still proudly manufactured in New Zealand – the Thermette name and brand is very much a Kiwi tradition.

Even the army is still using them – in fact the Thermette used to have an official UN equipment number. They aren't as widespread as they once were, because some modern armed vehicles (e.g. the New Zealand Army's light armoured vehicles or LAVs) have built-in water heaters, but the Armoured Corps used to routinely have one or two Thermettes in each vehicle. Boiling a Thermette used to be a great social event, attracting all the other military personnel in the area for a cuppa. Solders would fill a cut-down shell case with petrol and drop a match in to light it – about a cup will be enough to boil a standard-issue Thermette. However, this method did cause problems, particularly when it was necessary to boil several in a row – the unwary trooper filling a hot shell case for the first time might not notice the remnants of the previous fuel still boiling and consequently would be engulfed in flames as the vapour ignited. This looked particularly spectacular early in the morning or late in the evening when the air was still, but as Brigadier Sean Trengrove, director general of the Army Reserves, remembers it, the only injuries were to pride and eyebrows. The cuppa escaped unscathed.

Among other fans of the Thermette we can number Sam Neill, Kiwi actor, who remembers it from his youth in Central Otago. 'I learned how to stay downwind of the aromatic mānuka-fuelled Thermette to avoid the sandflies.'

The Thermette is a modest but brilliant invention, still given regularly as gifts and used fondly on family picnics, although driftwood is recommended as fuel, rather than the petrol method the army used! It's a brilliant piece of engineering – one of our real gifts to the world. Put the Thermette on and have a cuppa.

The World's Fastest Machine Gun

PUNKS NO LONGER FELT LUCKY

During World War II, when the Jerries and the Poms were fighting it out in France and the Japs were making their imperialistic way down the Pacific, the Allies were in need of a gun that could give them an advantage in the close-range combat that seemed to be the hallmark of the war in the Pacific. Whanganui-born Allan Mitchell, a ballistics engineer for the DSIR, gave them that advantage by inventing a series of machine guns with ever-increasing fire power. At the time the Thompson sub sten gun had a firing rate of 250–450 rounds per minute. Mitchell created a weapon that was light, easy to disassemble, cheap to make, and most importantly had a firing rate of over 1000 rounds per minute – that's 16 bullets a second. Mitchell was summoned to a war cabinet meeting to display the weapon, but when he pulled it out he caused quite a security scare. No wonder, when the weapon he had in his hand could have massacred the entire cabinet in seconds. Not content with this amazing weapon, Mitchell created his magnum opus – a machine gun to be mounted in aircraft that could fire over 6500 rounds per minute – 108 per second! At the time this was four times the speed of the then fastest gun, the Browning, and for many years was the fastest single-barrelled machine gun in the world.

The Chicken Switch

A POULTRY INVENTION

There is no doubt about it – farmer Robert Ellis (b. 1862, d. 1934) of Brightwater near Nelson is one of New Zealand's unsung geniuses. In 1910 he converted the waterwheel from a mill on his farm to power up his own house. The mill ground flour during the day, and generated electricity at night. In 1911 he decided to hook the wheel up to help power the local streetlights, but he didn't want to be bothered having to go and turn them on and off every night. Ellis, in a flash of inspiration perhaps only surpassed by the area's other genius, Rutherford (see page 114), realised that the hens running around near the mill might provide a solution. He hooked up a plank to the switch and put the plank in the henhouse. When the chooks came home each night and jumped on the plank to roost, they also turned on the lights in Brightwater and nearby Richmond. In the morning, when they got up to go about their business, the plank clicked up and the power went off again. No doubt there was also a roll of No. 8 Wire involved.

Homebake Heroin

NEW ZEALAND'S HOTTEST HOMEBAKERS

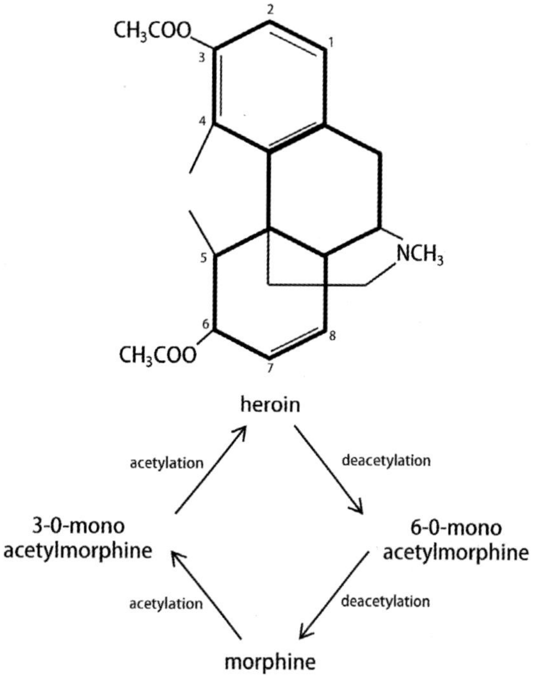

24

In 1974 New Zealand had just 24 criminal cases involving heroin.

Right: Kids, don't try this at home.

Just because it's illegal doesn't mean it isn't a classic, inventive, No. 8-Wire, bloke-in-a-shed response to New Zealand's isolation.

In the 1970s the world was experimenting with psychoactive drugs. In 1974 New Zealand had just 24 criminal cases involving the most addictive of them – heroin. By 1979 heroin was much more available in New Zealand thanks to extensive trafficking by syndicates like the infamous Mr Asia drug ring. Then the police orchestrated a crackdown on drugs. The Mr Asia ring collapsed and customs and the police strangled the flow of heroin into New Zealand. As the 1980s dawned the heroin addicts of New Zealand were starved of what they needed, and their necessity gave birth to a weird antipodean bastard – homebake.

Homebake is monoacetylmorphine – a sticky black or brown form of heroin which can be injected just as addicts would expect to inject the pure stuff derived from opium. But instead of exotic opium, sometime around 1983 enterprising and desperate New Zealand students of chemistry found a way to synthesise homebake from codeine or morphine tablets – both of which were available over the counter at the time in the form of medicines like Panadeine.

The homebake process was simple and required little special gear or training, but it was time-consuming and yielded only enough for a few 'hits'. Between 1983 and 1986 the New Zealand Police busted more than 90 homebake labs.

Our invention found its way as far as Australia, but hasn't had that much effect on the rest of the world. In most places opium is more readily available so homebaking isn't necessary. In some places they've developed other methods of synthesising heroin, but homebake is our very own.

Nowadays it's not so popular. Codeine is a lot harder to come by since a crackdown in availability of the standard ingredients in 2010 – well, a very Kiwi-style of crackdown where you can still buy them, but you might have to answer some very probing questions first! And the New Zealand drug scene has another and arguably worse scourge to deal with – 'P' (otherwise known as methamphetamine). At least we didn't invent that one.

The MONIAC

MODELLING THE ECONOMY WITH PIPES AND WATER

14

Moving beyond the rough prototype he'd built from old Lancaster bomber aeroplane parts, he made up to 14 water computers, and sold them to diverse places such as Harvard Business School, Istanbul University and the Ford Motor Company.

Right: Bill Phillips and his Monetary National Income Analogue Computer.

Understanding an entire economy – where the money goes, where the tax comes from, how investment and savings work – is hard. Everything is sort of interconnected, and changing one small thing can change the whole system in a complex and unpredictable way. That's why on the TV news most evenings, expert economists try to explain to us

This wasn't a normal computer (as we currently understand them), this was a machine that used fluid logic. Literally fluid – the machine used pipes and water.

laypeople why our mortgages just went up or why the price of big screen plasma displays is so good at the moment.

Alban William Housego Phillips – call him Bill (b. 1914, d. 1975) decided to make economics more accessible in 1949 when he built the 'Monetary National Income Analogue Computer' – also known by the admirable name 'Financephalograph'. Bill wasn't a normal economist, and this wasn't a normal computer (as we currently understand them), this was a machine that used fluid logic. Literally fluid – the machine used pipes and water, moving around in a series of tanks, to model the UK national economy. The water flowing represented money, so you could see money coming in from central treasury, and from other countries. The water flowed through the system between savings, investments and the like with some being syphoned off for taxes. The metaphorical alignment between the monetary system and the water-based machine continued when you could see savings dry up and debts overflowing.

Bill initially built the machine as a teaching aid, to show the workings of the UK economy, but an unexpected thing happened. He discovered his contraption had a very small error rate, and offered an effective way of modelling a very complex system. Long before we even knew what real computer models were, Phillips realised his MONIAC could be used as a simulator: put the right amounts of water into it in the right places, and the output from the computer was actually really useful in predicting the behaviour of the economy. Moving beyond the rough prototype he'd built from old Lancaster bomber aeroplane parts, he made up to 14 water computers, and sold them to diverse places such as Harvard Business School, Istanbul University and the Ford Motor Company. There is a MONIAC machine in our own Reserve Bank at the moment – ostensibly as a visual display of New Zealand ingenuity, but some allege it's still the computer model they are using.

Bill Housego Phillips – you have to use a middle name like that as often as possible! – was born near Dannevirke. He was quite an adventurer, spending time in China and the Soviet Union as a young man, working as a crocodile hunter and a gold miner, and eventually joining the RAF and fighting in Singapore. Phillips was captured and spent three and a half years as a prisoner of war in Indonesia. There he learned Chinese from other prisoners, and endeared himself to everyone by building secret water boilers to make tea. He powered his boilers by hooking them into the lighting system: the guards would wonder why the camp lights always dimmed around about eight each night. He also repaired and miniaturised a secret radio. After the war, he was awarded an OBE for his military service.

Phillips has been described as the Indiana Jones of economics and exemplifies the classic view of a Kiwi inventor – ingenious, individually gifted, insatiably curious and good with his hands.

Phillips has been described as the Indiana Jones of economics and exemplifies the classic view of a Kiwi inventor – ingenious, individually gifted, insatiably curious and good with his hands. But he was more than that too; he was a serious academic who contributed to the field of economics, through both the MONIAC and his Phillips Curve – an economic theory that describes the correlation between inflation and unemployment in an economy. Many thought if Phillips had lived longer (he died at the age of 60) he might have been in the running for a Nobel Prize for this theory. Indeed, a number of others who built their research on his publications received Nobel Prizes for their work. Phillips' research still influences macroeconomic theory today.

Phillips' is a life story full of adventure, invention, and contribution – not bad for an academic who described his own most famous theory as 'quick and dirty' and 'done in a weekend'. Add 'humble' to the list of his attributes.

7.

She'll Be Right

DEMOCRACY, THE GALAXY AND OUR GREAT-GREAT^{100} GRANDMOTHER

Careful observers will notice in this book an imbalance and an inequality in terms of the representation of the sexes. This is not due to any sexism on the authors' part, rather due to the dearth of female inventors and innovators. That is not to say there aren't any – just not many.

We have a theory on this. Many inventions are motor or mechanically orientated, and these areas of enterprise traditionally don't attract women. Another reason is that for much of New Zealand's history, women weren't afforded the same opportunities in terms of education and training that men had. Another reason may be shyness or an unwillingness to claim authorship of ideas.Yet another contributing factor may be that many inventions – probably 90 per cent – turn out to be a useless waste of time, and women are smart enough to avoid this wasted effort. Another possibility is that some of the inventions we attribute to males were actually invented by women, but for whatever reason, they weren't given due credit.

It's not just a New Zealand phenomenon either; this imbalance seems to exist worldwide. In 2010 women in the US held just 18 per cent of all patents, but the situation seems to be improving – 15 years earlier it was just 5 per cent. However, perhaps it's quality, rather than quantity, that brings equality. This chapter showcases some of the inventions by and for women in New Zealand.

Democracy

UNIVERSAL SUFFRAGE MEANS VOTES FOR ALL

1

On 19 September 1893, New Zealand was the first country in the world to have universal suffrage – and real democracy.

Right: Kate Sheppard, our glamorous suffragette, in 1905.

Sure, you may have learnt in school that the Greeks invented democracy, among other things – mathematics, law and philosophy spring to mind – but it's our contention that democracy in its full sense was invented by the Kiwis in 1893. For it wasn't until this date, anywhere in the world, that men and women could vote equally.

The precious privilege of the franchise has been hungrily guarded since the idea of democracy was invented in Athens around 500BC. For 25 centuries, the ideal of an equal say in government by all people always came with caveats – all people as long as you are a man, as long as you're not a slave, as long as you own land, as long as you are white. That we were first to remove all caveats means that, while democracy was certainly not invented by New Zealand, we were the first to perfect it.

The story of the fight for the vote for women in New Zealand is fascinating from our comparatively liberal position 100 years later. Many of the arguments and issues surrounding suffrage now seem quite silly and blatantly wrong – and that applies to issues on both sides of the debate.

Let's preface this topic with a bit of background on women's rights. Worldwide, women had been treated as second-class citizens basically forever. Indeed, the rights of women in most of the Western world were, until the mid-1850s, on a par with those of lunatics, prisoners and children. New Zealand – in no small part due to our remoteness from the influence of Mother England – was slightly more liberal. In New Zealand, women in the late 1800s had a chance to get an education. The first woman in the British Empire to complete a tertiary degree, Kate Edger, graduated from the University of New Zealand in 1877. Yet there still existed the lamentable situation that, once every three years when the government of the country was being chosen, women's opinions were ignored.

The suffragist movement worldwide had been under way for many years, and it wasn't just women who were asking that their voices be heard. The

British philosopher and economist John Stuart Mill presented a petition to the British House of Commons in 1867 calling for the vote for women, but it was ignored, and Mill lost his own seat the next year.

In New Zealand the suffragist movement was, for better or worse, entangled with the issue of temperance. Many feared the establishment of voting rights for women was a vote for the prohibition of alcohol, a view that was not helped by the suffragettes themselves, who organised themselves under the banner of the Women's Christian Temperance Union (WCTU). The main concern of the WCTU was that excessive use of alcohol was undermining the family unit, and further, that the woman's work for the economic well-being of the family was being frittered away by dad drinking all his pay.

Of course, the idea of giving women the vote wasn't limited to this one issue. The WCTU campaigned for equal divorce laws, raising the age of consent for sexual intercourse from the prevailing 12 years, and preschool education. They were also vocal in their opposition to the wearing of corsets, which they saw as symbolising the restriction of women.

Katherine Malcolm (b. 1847, d. 1934) – more familiar to us today by her married name, Kate Sheppard – was one of the leaders of the WCTU and, therefore, of the suffragette movement. She was born in England and moved here as a young woman. She was regarded as a highly intelligent and well-educated woman. She also had a supportive husband who gave her the encouragement, opportunity and financial means to travel the country expounding her views on women's rights.

Needless to say there were those for whom the idea of women having their say did not hold any joy, and these weren't just necessarily the men who enjoyed excesses of alcohol or liked the look of women in corsets. Universal suffrage had one of its biggest opponents in Richard John ('Dick') Seddon, an influential politician whom Sheppard and the WCTU would need to defeat if their cause was to triumph. Between 1887 and 1893 the WCTU presented to Parliament three increasingly large petitions. Indeed, a bill was passed by Parliament in 1891 supporting women's right to vote, but this bill was subsequently thrown out by the legislative council (the upper house).

In 1893 the suffragettes' best chance came. Another bill was passed through the lower house, and it looked like it might head through the legislative council also; but then came a blow to the movement. Then premier John Ballance, a supporter of women's rights, died, and Richard Seddon became the premier. He tried to use his influence to change the votes of some members of the legislative council, a plan that backfired on him. News of his underhand approach reached other members of the council, who changed their votes accordingly.

The bill was passed. But it still wasn't law until the Queen's representative and governor, Lord Glasgow, signed it off. Some efforts were made by opponents of the law to stop him, and in their last symbolic protest, the suffragettes sent camellias to all the members of Parliament – white ones to their supporters, red to their opponents. Lord Glasgow signed, and on 19 September 1893, New Zealand was the first country in the world to have universal suffrage – and real democracy.

Our idea to have 100 per cent pure democracy did catch on in the rest of the world, but it took them a while. It wasn't until 1906 that Finland became the next nation to embrace universal suffrage. Australia blotted their record by awarding women the vote as early as 1902, but not achieving universal suffrage until a full 60 years later when indigenous Australians were finally allowed a say. But last place has to go to Switzerland and Saudi Arabia. The Swiss didn't award women full voting rights until 1990 (that is not a typo) and Saudi Arabia will be holding out until 2015.

Kate Sheppard on her victory for human rights:

"It does not seem a great thing to be thankful for, that the gentlemen who confirm the laws which render women liable to taxation and penal servitude have declared us to be 'persons'... We are glad and proud to think that even in so conservative a body as the Legislative Council there is a majority of men who are guided by the principles of reason and justice, who desire to see their womenkind treated as reasonable beings, and who have triumphed over prejudice, narrow-mindedness and selfishness."

The Eve Hypothesis

MY MUM KNOWS YOUR MUM

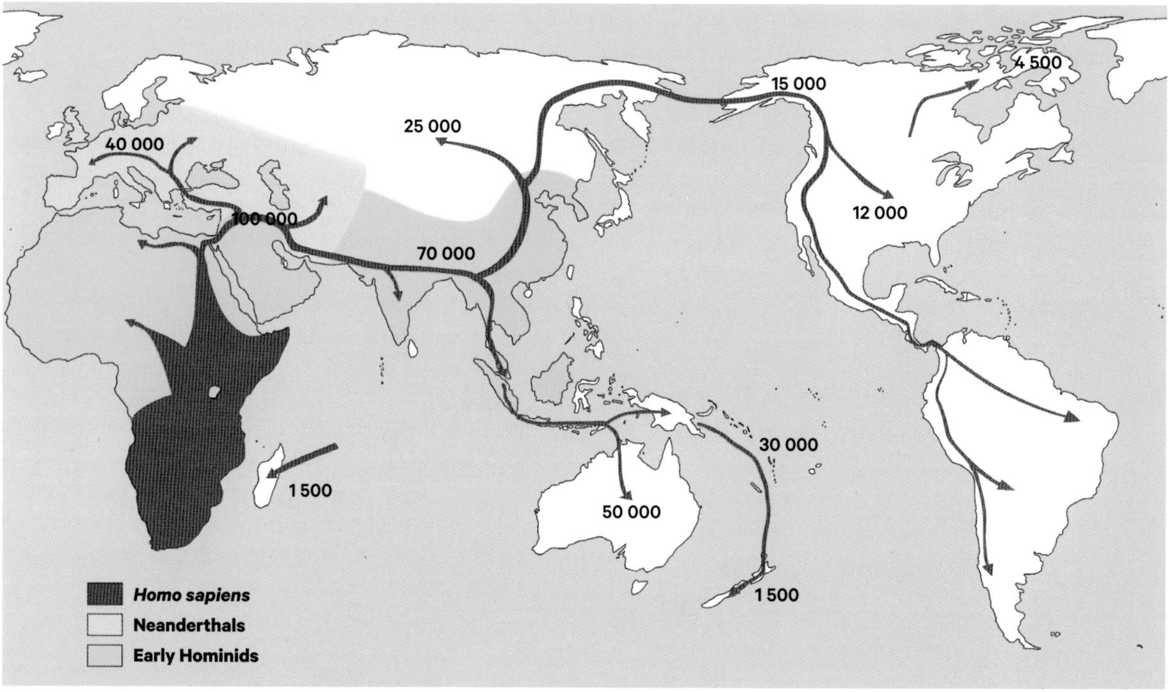

Map of early human migrations (measured in years) based on the 'Out of Africa' theory. New Zealand was one of the very last land masses to be settled.

Professor Allan Charles Wilson (b. 1934, d. 1991) was a scientist and evolutionary biologist who hailed from New Zealand. He had a profound impact on how we humans think about our place in the animal kingdom, and yet is largely unknown in his home country. It would be easy to mistake Wilson's work as a sort of anti-evolution or pro-creationist pseudo-science, but quite the opposite – Wilson was a serious scientist. He showed that humans are not so different from apes, and that all humans are much more closely related to each other than had previously been thought. Wilson's theories are now widely accepted, but at the time he proposed them, they were revolutionary (or perhaps, re-evolutionary?).

Wilson was born in Ngaruawahia and educated at King's College in Auckland, and gained his bachelor's degree from the University of Otago. Then, like many of our brightest scientists, he left our shores to continue his study in America. Ending up at the Berkeley campus of the University of California in the late fifties must have sparked all sorts of ideas in a young man's head, particularly one with a good working knowledge of biochemistry, and it didn't

200,000

Their startling conclusion was that all humans were descended from one single female who lived about 200,000 years ago in northern Africa.

Left: Dr Allan Wilson.

take long for Wilson to publish his first very controversial theory.

Wilson gained world's attention in 1967 when he developed a new technique to measure the "immunological distance" between species. This technique compares cell proteins in different species to tell how closely related they are, and thus show the relative age of species. Wilson dubbed it the 'molecular clock'. It was a brand new way of looking back at the history of species and it immediately gave results that surprised the world, showing that humankind was in fact much younger than had previously been thought. Conventional fossil dating process had the date of the start of human evolution at about 20 million years ago, but Wilson's technique showed that our genetic material was very similar to that of the primates. Wilson deduced that mankind had branched off from the other primates only 5 million years ago. This 'recent-origin' theory was a complete shift in thinking from the prevailing views, and remained controversial until it was confirmed when anthropologists discovered the Lucy fossils in 1974. For this stunning piece of work alone, Wilson would have been celebrated.

As viewers of *Forrest Gump* will attest, the late 1960s at Berkeley were amongst the most turbulent, yet creative, times that we have known in modern history – a complex backdrop to Wilson's controversial work. Although his theories were backed by as much evidence as any of his contemporaries', his work was the subject of vitriolic attack from ill-informed critics. As a quiet and humble person, he was sure his findings would be accepted as accurate, which indeed they were. Slowly he and his ideas became accepted, and he gained prominence in his field.

Then, in the early 1980s, he dropped another bombshell. Wilson had worked out exactly when, and where, the human race had diverged. He and his team studied the diversity of a particular type of DNA only inherited from females, 'mitochondrial DNA', from people of many different ethnic backgrounds worldwide. Their startling conclusion was that that all humans were descended from one single female, about 200,000 years ago, in Africa.

Wilson himself wasn't 'anti-evolution' but his findings were certainly interpreted that way – the results of this announcement were loud and varied. *Time* magazine dubbed it the 'Black Eve Theory'. *Newsweek's* issue on the theory sold a record number of copies. The media saw it as simple-minded creationism, but the religious didn't like it much either. Much of the controversy was because Wilson's theory was simplified by the media and widely misunderstood. The theory isn't that Wilson's so-called Mitochondrial Eve was the only woman alive in her time, just that we all descended from her.

The 'Eve hypothesis', with a few improvements on the dates, is now the generally accepted theory of human evolution. It has had huge ramifications for all of anthropology and is still causing controversy today. Unfortunately Wilson can no longer defend it, as he died at the age of only 56, but his name adorns one of New Zealand's Centres of Research Excellence, and he left us with one of the best steps towards finding our origins in, oh, about 200,000 years. Professor Allan Wilson remains the only New Zealander to win the American MacArthur Fellowship (the 'Genius Grant'). He was one of New Zealand's most significant scientists – and that is saying something.

Baeyertz Tape

ACCURATELY ESTIMATING BIRTH DATES

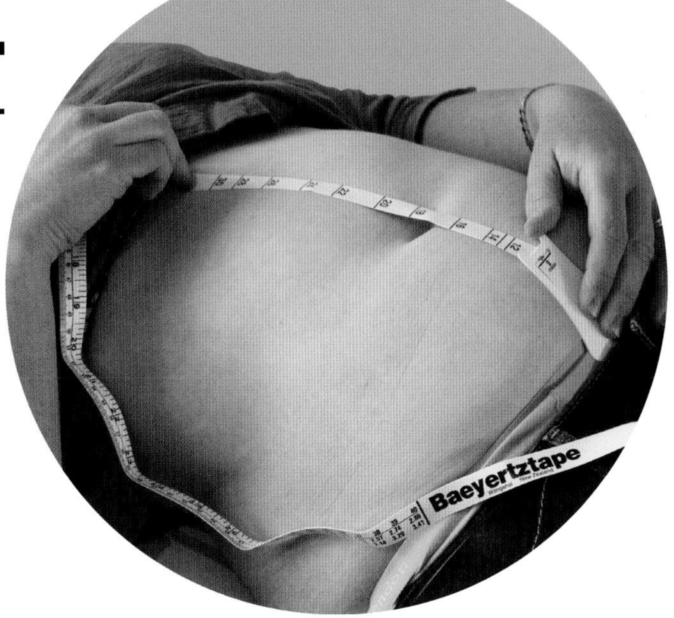

95

95 per cent of all pregnancies last between 261.3 and 270.7 days – roughly 38 weeks and four days.

Right: Baeyertz's tape is still used by midwives today. Not normally on themselves, though.

You may not know this, but if you go up to a pregnant woman and measure the distance from her pelvic bone to the top of the fetus, you will: a) know how far through her pregnancy she is; and b) get a slap. If you do it enough times and plead that it's for scientific purposes, then you are Dr John Baeyertz (b. 1924, d. 1998), an obstetrician and gynaecologist from Whanganui.

Dr B (as we'll call him, to lessen the chances of misspelling his surname) realised that the traditional methods of calculating a baby's due date – either by calculating from the date of the last menstrual period, or by the date of conception – were both error-prone and unreliable. He had of course heard of the technique of using the measurement from the symphysis to the fundus – doctor-speak for 'pubes to the top of the baby' – but there were problems with this technique too. No one had come up with a way to accurately use the measurement and most settled for roughly one centimetre per week of gestation. Compounding this, the invention of ultrasound technology gave hospital doctors very accurate information, leaving GPs out in the cold and on their own.

Dr B realised that with the advent of artificial insemination he could accurately measure the time between conception and delivery, and formulate a precise technique to be used for all women. He spent over 13 years collating the data from 127 pregnancies, including some twins, some abnormal pregnancies, some women who were early, some who were late... From his studies he gained a wide cross-section of pregnancy in New Zealand – or at least, in Whanganui. He took these results and expected that they would apply to the greater community.

Thanks to his studies, he found that a human pregnancy lasts 266 days with a standard deviation of 4.7 days; that is to say 95 per cent of all pregnancies last between 261.3 and 270.7 days – roughly 38 weeks and four days.

This was one of the most accurate measurements of pregnancy ever taken. It gave Dr B the data needed to create a measuring tape calibrated in weeks, which could be used by physicians everywhere for predicting birth dates. The Baeyertz tape was patented in 1982 and is now sold worldwide, bringing Dr B (hopefully) fortune and fame, albeit posthumously.

Biodegradable Hair 'Not-foils'

STREAKS AHEAD OF THE COMPETITION

It's important (for the blokes reading this) that we explain what foils are. Basically, when your wife goes to the hairdresser and spends four times as long and six times as much money as you do, it's probably because of foils. Literally they are bits of tinfoil, used when colouring hair to give it streaks. They act as a palette: you lie the strands on the foil, paint on the dye then wrap it up to keep the different strands separated. You'll be further interested to know that the sheets of foil get thrown away, and given most women look like a metallic Medusa when getting foils, it causes an enormous amount of wasted tinfoil from every salon.

Like most in her trade, Dunedin hairdresser Amanda Buckingham spends a lot of time standing and contemplating things, and one day had a revelation – what if she could replace the foil with something more sustainable? After two years of development with the help of the local polytech, she now has a biodegradable product made of recycled industrial waste, which she can wash and reuse. Customers also like it because it doesn't scratch and rustle when being used. Buckingham's One Systema product is now being distributed across Australia and New Zealand, and even Italy, with wholesalers and distributors lining up to get access to her product. While the name of the product looks like the English word 'One', it's actually that Māori word 'One', which means sand and earth, and with the interest they are currently getting, Buckingham is sure to stay streaks ahead of the competition.

After two years of development with the help of the local polytech, she now has a biodegradable product made of recycled industrial waste, which she can wash and reuse.

Reading Recovery

TEACHING THE WORLD TO READ

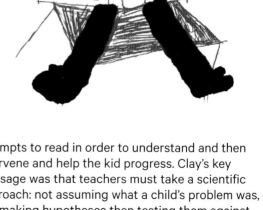

80

Eighty per cent of children who start the Reading Recovery completely illiterate are able to catch up to their classmates in 12 to 20 weeks, achieving five times the average rate of reading progress.

If you are reading this now, there's a fair chance you were helped by Dame Marie Clay's invention. Clay (b. 1926, d. 2006) was born in Wellington and got her first university degree from the University of New Zealand (until 1961 all the major tertiary institutions here were amalgamated and gave degrees under one name, making it seem embarrassingly like we had only one university).

Clay made careful study of the way children learn to read through her master's degree, PhD and afterwards. In 1960 she began teaching at the University of Auckland where in 1975 she became New Zealand's first female professor.

In 1976 Clay began to turn her research into a tool to help children who struggle with learning to read. In the first year of school, some children fall behind immediately with their reading and writing. Clay formed a plan called Reading Recovery to intervene early and bring struggling little readers up to scratch with their peers. Reading Recovery is based on daily 30-minute one-on-one sessions between specially trained teachers and children. The teachers are trained to observe, analyse and interpret the child's attempts to read in order to understand and then intervene and help the kid progress. Clay's key message was that teachers must take a scientific approach: not assuming what a child's problem was, but making hypotheses then testing them against a child's actual responses.

The programme was strikingly successful. It was rolled out to all New Zealand schools, then taken up in the UK and US. It is now taught in all English-speaking countries and has been translated to Spanish and French. Eighty per cent of children who start the Reading Recovery completely illiterate are able to catch up to their classmates in 12 to 20 weeks, achieving five times the average rate of reading progress. Clay's programme has faced criticism because it isn't cheap to implement, but it's taught in over 13,000 schools in the United States alone and has helped millions to learn to read.

Clay was made a Dame in 1992, and in 2003 was voted most influential person in the field of literacy by the National Reading Conference of America. If you haven't understood any of this then perhaps you should undertake Reading Recovery immediately.

Flower-vending Machine

FLOWERS AT ALL HOURS

2013

The first kiosk opened in Blackfriars Station in London in 2013, and immediately all along the train route marriages improved.

Right: A fully automated floral feast.

If the stamp-vending machine (see page 122) is one of New Zealand's oldest and most successful inventions, the flower-vending machine is one of our newest. Whether it is successful, only time will tell. Famous Kiwi film production designer Andrew McAlpine (*The Piano, Sid and Nancy, The Beach* and more) has designed and built an unmanned kiosk that automatically vends specially crafted bouquets of flowers 24/7. Using the kiosk's screen you choose a bunch and pay by credit card, then the door opens and you take your flowers. The first kiosk opened in Blackfriars Station in London in 2013, and immediately all along the train route marriages improved.

The Origin of the Cosmos

NEW ZEALAND'S OWN GALACTIC MIDWIFE

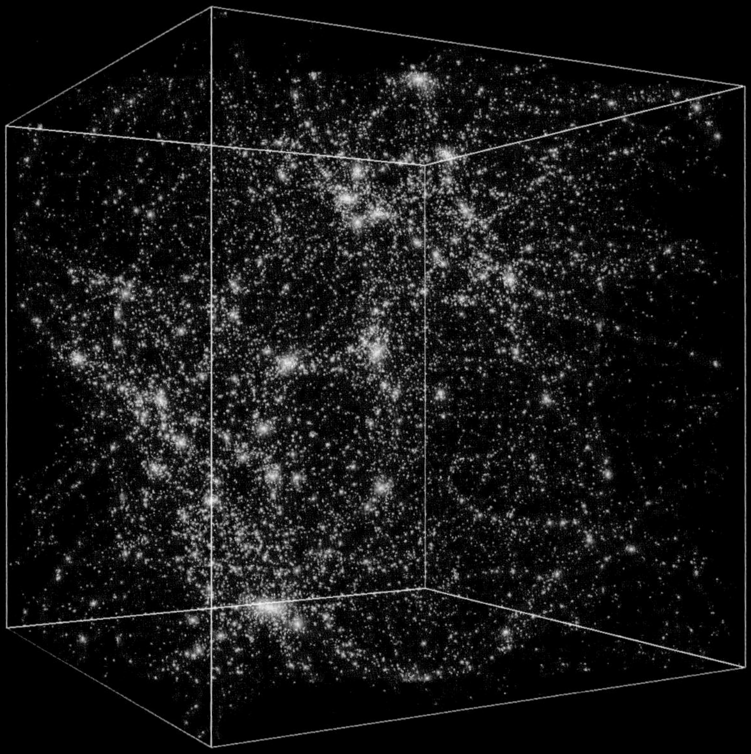

Beatrice Tinsley (née Hill; b. 1941 d. 1981) was born in Chester, England, but moved to New Plymouth after the war. She attended New Plymouth Girls' High School, with my mother-in-law who along with her classmates mates called her 'Beetle'. Not in a derogatory way, they claim, but in reference to her name. Her dad was also mayor of New Plymouth for a while, by the way.

Beetle went on to study at the University of Canterbury, where she decided she would become an astrophysicist – not a decision that comes lightly, I'm sure. After getting their degrees, she and husband Brian moved to Dallas, Texas, to continue their studies. Beatrice gained a PhD with flying colours, getting marks of 99 and 100 per cent, and then set about making a name for herself in her

1986

This award, available since 1986, is the only major award honouring a woman scientist to be created by an American scientific society.

Left: Her classmates called her 'Beetle' – Dr Beatrice Tinsley.

Throughout her life Tinsley authored over 100 scientific papers; she was heralded as a great scientist and teacher, and an inspiration to women scientists both in America and New Zealand.

chosen area of study, the evolution of galaxies.

On the total scale of things, a galaxy is the biggest known 'thing'. The current model of the universe has it being made up of hundreds of millions of island galaxies, where a galaxy is a collection of hundreds of millions of stars, some with solar systems like our own.

Tinsley's work was on the development of these galaxies and of the stars within. She asked, 'How did they form?' She created models of galactic formation that were said to be more realistic than other models at the time, combining a detailed understanding of stellar evolution with knowledge of the motions of stars and nuclear physics. In short, Tinsley married together many branches of knowledge and created a workable model of galactic creation.

Her work had a profound effect on the study of astrophysics at the time, changing the direction of thinking on galactic formation. Tinsley herself went on to become a professor of astrophysics at Yale University in 1978, but sadly she was diagnosed with cancer the same year, and passed away three years later; she continued to work right up until her death.

Throughout her life Tinsley authored over 100 scientific papers; she was heralded as a great scientist and teacher, and an inspiration to women scientists both in America and New Zealand.

In her honour, the American Astronomical Society established the Beatrice M Tinsley Prize for outstanding creative contributions to astronomy or astrophysics. This award, available since 1986, is the only major award honouring a woman scientist to be created by an American scientific society. Somewhat sadly, though, Tinsley was also a victim of her gender. After her marriage in Christchurch, she was barred from working at the university while her husband was employed there. This was then repeated in the US when she tried to work at the University of Texas at Dallas, and again was not allowed to work, in part because her husband worked there. Shame on you, 1960s academia.

Tinsley is now rightly recognised as one of New Zealand's most important scientists. Her work created a number of awards and endowments, but perhaps more permanently, she has a mountain named after her in Fiordland (Mt Tinsley at 1537m is in the Kepler Mountains – Kepler was also an astronomer) and even an asteroid named after her, *3087 Beatrice Tinsley* (1981 QJ1).

The Freezer Vacuum Pump

RONGOTEA'S GIFT TO THE WORLD

Right: The iconic freezer vacuum pump in action.

If anybody is New Zealand's kitchen whiz, it's Norma McCulloch. Sure, you've got your Alison Holsts and your Annabel Langbeins – but all they've given the world are date muffins and saffron spinach risottos. McCulloch (b. 1933, d. 2010), on the other hand, gave the kitchen world some indispensable inventions.

In 1968, Rongotea home economist Norma McCulloch was searching for a better way to avoid freezer burn. Not for herself, you understand, but for her silverbeet, lamb shanks, cabbages and kings. The basic theory is that food keeps much better in the freezer if all the air has been sucked out of the surrounding bag first. Previously, housewives had been squeezing the air out of the bags by hand, or even, heaven forbid, sucking it out with a straw. Thus, the freezer vacuum pump was born.

My mum had one and so, probably, did yours. A cardboard (later plastic) tube, with a metal inner tube you pumped up and down vigorously. This product quickly became the cornerstone of a multinational company – McCulloch Industries Ltd and sold around the world. Norma's son Richard came on board and the company grew, developing a range of products including, somewhat ironically, a pump for getting air into balloons and lilos. It's unclear whether the company ever went the final step and created an invention for taking air directly from freezer bags and putting it into lilos, but surely this idea was toyed with.

Not content with merely creating a range of kitchen products, the McCulloch family formed McCulloch Medical, and Norma invented a range of resuscitators and face masks – providing air to those who need it urgently. These were used for years by the Australian Defence Force and the United Nations. The Breath of Life resuscitator was designed for use with humans, but has spawned similar inventions to resuscitate animals like piglets and alpacas, and Richard McCulloch now markets the Next Step Neonatal Resuscitator. This is a family that really likes moving air around.

Demand for the original freezer vacuum pump has died off over the years because of the availability of freezer bags and the like. But in 2003 Norma McCulloch was named as one of the top 10 female inventors in the world at the Global Women's Innovator and Inventor Awards.

Norma and her family's diverse products go to show that a good idea coupled with good execution can go a long way to global success and save a lot of silverbeet along the way.

Asthma Drug

A SINGULAIRLY (SORRY) GOOD DRUG

Asthma is a terrible affliction. New Zealand has an appalling asthma record, with the disease killing around 130 New Zealanders per year, costing the health system an estimated $375 million.

In 1998 a team from the drug company Merck & Co. announced a breakthrough new drug in asthma treatment called 'Singulair'. A key player in the team that created the drug was Dr Jilly Evans. Evans was born in Napier, and has a master's degree in cell biology from the University of Auckland and a PhD from the University of British Columbia. What Evans and her team came up with was a chewable tablet, to be taken once a day, rather than the inhaler-type asthma drugs we're all used to – a preferable option for everyone, particularly children. The drug works by targeting and blocking the effects of leukotrienes, which cause inflammation, airflow obstruction and secretion of mucus, so the asthma effects never occur. This is different from conventional medicines that work to relieve the symptoms of asthma. Singulair has been proven to decrease asthma attacks significantly, and has helped wean many people off the old treatment of inhaled steroids. It's been described as the biggest step in the treatment of asthma for over 20 years and has been a multi-billion dollar winner for Merck.

Breast Protectors

PROTECTING THE GIRLS

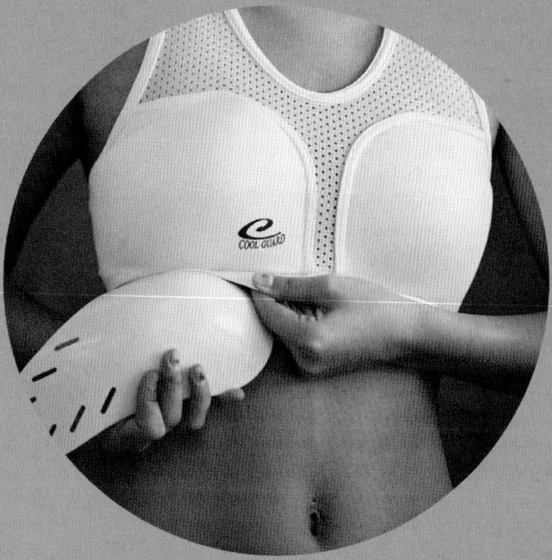

In 1981 Ces Richie, Wynn Martin and Max Rutherford were asked by a nun to help design a way to protect the breasts of the girls playing contact sport at her school. They lived in the shadow of Mt Taranaki, whose soft, molten interior is protected by a hard shell-like exterior, and no doubt drew inspiration from the conical volcano. Their company QP Sport made a fibreglass mock-up of a chest protector which was held on by bra straps and could be worn under the girls' soccer uniforms. They gained a New Zealand patent for the protector and developed a range of revolutionary bras – for example, their CoolGuard product has two plastic guards that are inserted into a specially designed sports bra. The guards are made of unbreakable, Tupperware-style polypropylene and are moulded to spread impact. They are very popular in rugby, in combat sports like karate and boxing, and even in paintball and roller derby. QP Sport's products are sold under a number of brands in over 50 countries – and all thanks to that nun.

Marching Girls

IT'S GIRLS MARCHING

If it was the Great Depression and people in New Zealand had no money to spend on leisure activities, could a pastime be invented that would take girls' minds off their troubles, cost nothing and be great fun? Could it be a pastime that would give women a team sport for the summer? Could it be one of the only sports that has ever been invented in New Zealand, to be competed in internationally? The answers are all 'yes', and the sport was marching.

Marching associations were grateful to pipe and brass bands for giving them music to compete to, and the bands were grateful to the marchers for giving them a reason to play marches that nobody else really wanted to hear.

This first record of marching girls is in the official

history of the visit of the Duke and Duchess of Cornwall and York in 1901. During the Dunedin celebrations, about 500 girls marched past as a celebration of 'girls' drill'. By the late 1920s, marching was established in Otago as a sport. Teams of nine girls were organised and drilled – originally mainly by military men. The marching was based on the *Army Manual of Elementary Drill*, but instead of men calling out orders, the girls took their cues from music. By 1933 business-house teams from factories, hospitals and the armed forces were competing in championships, and in 1945 a group of businessmen in Whanganui met to form a national body to organise competitions between clubs. Because it was cheap and needed no special equipment or playing fields, the sport developed rapidly.

All over the country young women drilled and drilled, were outfitted into uniforms and competed against rival teams. Teams are judged in competition by deducting points for the most slight and subtle errors: too high an arm swing here, slight misalignment of a head there. Marching associations were grateful to pipe and brass bands for giving them music to compete to, and the bands were grateful to the marchers for giving them a reason to play marches that nobody else really wanted to hear.

It may seem unusual that there are thousands of young New Zealand women keeping large hairy hats in their cupboards, but much more stupid sports have become Olympic events. Synchronised swimming, for example, hasn't even got a good name! How much fun would marching sound if it was called synchronised walking? The tall busby-style hats they wear are a recent addition to the uniforms, and by no means mandatory. Throughout the early years of marching in New Zealand they wore more military-style hats, often with a fan of feathers at the front.

It wasn't long before New Zealand marching began to conquer the world. In the early 1950s, several international showcase tours were carried out, resulting in an Australian Girls' Marching Association. In the 1970s and '80s, an annual competition was held between New Zealand and Australia. This 'Bledisloe Cup of Marching' has fallen away now as Australia is struggling to get the numbers along and won't agree on the rules.

Perhaps the international high points for New Zealand marching were the times the Lochiel Marching Club of Wellington was invited to the Edinburgh Military Tattoo. Another high point might have served as a warning to the organisers of the sport. In 1993 marching was celebrated on a New Zealand stamp issued as part of a series depicting New Zealand life in the 1940s. If marching had a problem with its image of being something from the past, it began to show in the numbers of competitors, which fell during the 1980s and '90s.

But in 1997 and '98 the organisers did something about it. Working with the Hillary Commission, they started by reorganising the structure and changing their name – from the New Zealand Marching Association to Marching New Zealand. Thanks to this, and a few other more tangible changes, the numbers are starting to come back. They are fighting off competition from cheerleading and other such 'Jilly-come-lately' sports and keeping up with the times. Marching New Zealand president Pam Findlay says it's all about striking a balance between the needs of contemporary youth who want excitement and fun, and more traditional coaches and administrators who want the pompous ceremonial music of old. Events are shorter now with more variety, and they allow the girls to choose their own music more, although thankfully twerking has not yet been accepted as a recognised move.

500

This first record of marching girls is in the official history of the visit of the Duke and Duchess of Cornwall and York in 1901. During the Dunedin celebrations, about 500 girls marched past as a celebration of 'girls' drill'.

Left: Champion Wellington team the Sargettes march in their Highland-inspired uniforms, 1951.

Lucy's Agar Mission

SEAWEED FOR THE WAR EFFORT

1941

In 1941 the sole supplier of agar to the world was Japan. Can you see the problem?

Right: Dr Lucy Moore, discoverer of a new technique for making agar.

Now I'll bet you haven't given agar a lot of thought. But this unassuming gelatinous substance is vitally important in many endeavours, not the least of which are science and…ummm…meat-canning.

Agar is made from seaweed, boiled down and separated off. The resultant substance is jelly-like and clear, and looks a lot like gelatine, which is extracted from animals. Indeed, you could say agar is a superior-quality vegetarian gelatine. Agar is that stuff you see on top of Jellimeat to stop it going bad; it is used in DNA fingerprinting technology; it is used in the soft-centred chocolates Cadbury make, and, perhaps the one we all know from school, it is used in scientific laboratories to grow cultures of bacteria. The little plastic dishes of agar, with their distinctive smell, are the trademark symbol of laboratories the world over.

In 1941 the sole supplier of agar to the world was Japan. Can you see the problem?

Of course, Japan's entry into the war meant New Zealand's agar supply dried up overnight, leaving all sorts of industries in dire straits. The hunt was on

Above: Schoolchildren from around the coastline of New Zealand were enlisted to help.

She offered a small amount of money to local schools in return for their students clambering around rocks collecting soggy bits of seaweed.

for a new source of agar, and fast. And it fell to Kiwi botanist Dr Lucy Moore (b.1906, d.1987) to provide the solution.

Aged 35 when the war broke out, Moore – with extensive training in marine botany that meant she was an expert on seaweeds – was the ideal person to find a weed that could not only provide a source for agar, but also provide it in sufficient quantity to fill all the needs of the nation at war. She was immediately dispatched on a search along the coastlines of the country to find a New Zealand seaweed suitable to make agar.

Moore spent months searching until she found a likely candidate. *Pterocladia lucida* is a seaweed found in abundance around the Bay of Plenty and on the East Coast of the North Island. Moore found that, when boiled down and separated, the weed proved a source of excellent quality agar – even better than the Japanese agar, in fact. But now the problem was getting enough of it.

Again Moore came up with an ingenious idea. She offered a small amount of money to local schools in return for their students clambering around rocks collecting soggy bits of seaweed. Indeed, when presented with something practical they could do for the war effort, the entire community rallied around and soon seaweed drying on farm fences was a common sight right along the East Coast. The wartime collection helped the country, and it also helped the local school. It was the 1940s equivalent of our collecting phone company talking points for our local school, or the 1970s effort of collecting coupons from tea packets.

As it turned out, Moore's work not only assisted New Zealand in times of war, but her discovery of an entirely new source of agar literally on our very shores developed a new New Zealand industry that continues today. While it is a small country, New Zealand's coastline is bigger than China's and almost as long as that of the US – we have the ninth longest coastline of all countries – which means we have ample opportunity to harvest from the sea.

Scientists Ian Millar and Richard Furneaux did a great deal of work on the native seaweed *Curdiea,* which can stay gelatinous to about 105°C, making it very useful in the food industry. One company, Coast Biologicals, processes hundreds of tonnes of seaweed into agar each year, and with agar fetching about $100 per kilogram on the international market, it's a multi-million dollar business. Thank you, Dr Moore!

Kindling Cracker

FORGET 'BLOKES IN SHEDS'

10

Kindling Crackers have sold online to more than 10 countries.

Right: Ayla Hutchinson is from a long line of inventors.

Ayla Hutchinson (b. 1999) was not so much a 'bloke in shed' as a 'teenage girl in shed' when she invented this – a device to safely cut kindling wood from larger pieces. Ayla came up with the idea for the wood chopper after seeing her mother clip the end of her finger with the hatchet while chopping wood.

She took her design to her father, who welded the head of an axe into a small enclosure. The wood is then put on top of the axe blade and can be hammered down, splitting the wood as it goes. A simple idea, literally turning the conventional way of chopping wood on its head. It means the blade itself never moves, so you'd have to be pretty unco to cut yourself.

While the invention itself is clever, so was the way she and her family then went about capitalising on it – she entered (and won) an innovation award at the 2013 Fieldays event, another at the New Zealand Innovators Awards and a handful of others to boot. The publicity from these helped created demand, and with a provisional patent safely filed, the family is now busy commercialising and selling the Kindling Crackers to both the New Zealand and global markets. Dad Vaughan has pared back his time in his other businesses to spend time on this product. They're in that difficult stage where they know they have a great idea and a great product, and they just need to get the marketing and sales humming so they can start to drive real volumes. They've gone about it the right way, though, selling online to over 10 countries. Orders are starting to pick up and they hope to see good demand for their New Zealand-made product – there is even talk of a licensing deal with a major US company to take the product global. And when she's not at school, Ayla herself is helping produce, paint, pack and market the Kindling Cracker.

Ayla comes from good innovative stock – not only were her father and grandfather keen engineers and innovators; her great-great-great-grand uncle George Fell Hutchinson was a well-known inventor in Hawera where they lived, creating a hydraulic water ram; and her great-grandfather William Hutchinson created the revolutionary Hutchinson Individual Milker ('Watch your cows' udders become soft and pliable . . .').

Ayla herself has her eyes set on a career as an innovator, and with the burgeoning global success of her first product, she's already off to a cracking start.

Powdered Paint Pigments

BETTER THAN POWERED PIG PAINTING

Rachel Lacy has paint in her veins. Her family owned the Aalto paint stores around New Zealand and Australia, which may explain why Lacy had an epiphany about the paint industry. What she was good at was colour, not paint. Furthermore consumers are mostly interested in colour these days, rather than paint. But with the current method of delivering paint to people there is no way to separate the two. Lacy decided to change that.

Her idea was simple: manufacture powdered colours that can be mixed with a base easily and at home. The world is full of beautiful colours. Designers such as Ralph Lauren and even architects like Le Corbusier have come up with palettes of colours. But they aren't available to everyday consumers with current technology. They can't necessarily be tinted in your local paint shop, and paint is very hard to order from overseas.

The way we retail paint hasn't changed since the introduction of the tinting wheel into the paint shops in the 1960s. It's messy and inaccurate, and you can never get exactly the same colour twice. Lacy's vision was a world of powdered colours available online to everybody, easily posted around the world. You can buy a base at the paint shop, then order whatever pigments you like from anywhere. The equally important corollary of her vision was that colour designers around the world (Lacy's future partners) will have access to a global market for their colour collections – and all through a new way of delivering them.

Could an idea as simple as this be new? Lacy didn't know so she ordered worldwide research from expert patent lawyers and technology experts. Says Lacy, 'We looked under every rock to make sure nobody else was doing it.' They weren't. The market was wide open and the lawyers confirmed that this was 'brand new, disruptive technology'.

Figuring out how to make the powdered colours was a 'massive technical challenge'. Partnering with chemist Dr Simon Hinkley at Callaghan Innovation, Lacy spent 18 months in a small lab set up in Auckland, perfecting the technology to produce brilliant-coloured powders that can be instantly mixed with a paint base. The product is called drikolor.

In the last few years things have gone extremely well. Attempting, in 2012, to raise $600,000 from angel investors through Icehouse, Lacy raised $700,000. The R&D to commercialise the technology has been successful. Then in late 2013 drikolor beat out 400 other start-up companies in a Hong Kong competition to identify outstanding commercial potential. At the time of writing, Lacy is eagerly awaiting a massive drying kiln being shipped from Asia – the next step in bringing the world a new way to buy colour.

Lacy sees New Zealand as the perfect place to breed her invention. Because we're small she's had to deal with all aspects of the paint industry – from colour design to paint manufacture to retailing. This broad vision is what enabled her first to imagine a new way to do things, and then gave her the capabilities to pull it off. What's the ultimate aim? 'We're going to be the Apple of the paint world. We're going to change the way the world colours paint in the same way Apple changed the way the world listens to music.'

8.

The Three Rs

RUGBY, **R**ACING, BEE**R**
(AND OTHER DIVE**R**SIONS)

In this chapter, we look at contributions to recreation and the holy trinity of rugby, racing and beer. Kiwis missed the opportunity to invent beer by about 6000 years, but my word we improved its creation. Given enough time, we would have thought up rugby, but instead we became the best at it then made the world come up to our standards. We didn't invent horse racing, but we have had some of the world's greatest horses and revolutionised racecourse betting with the automatic tote machine.

Relatively few of the pastimes we get involved in are our own – perhaps we're too practical a nation to spend time inventing things that won't put food on the table. But we certainly aren't afraid to spend time getting good at the games that others invent. The incredibly close results of the 2011 Rugby World Cup and the 2013 America's Cup showed that the nation now has the ability to win, or lose, not only at sports that celebrate fitness or a well-formed scrum, but also those that draw on science, engineering, technology in a sophisticated yacht design. That we can talk almost as fluently about a carbon-fibre hull as we do about a hamstring injury shows that even in the world of games we are happy to step into a new world.

Continuous Fermentation

THE GIFT THAT KEEPS ON GIVING

15

William died in 1918 of the flu, so Morton, aged only 15, took over the business. Talk about a recipe for disaster – would you leave *your* teenager in charge of a brewery?

Right: The WilliamsWarn personal brewery – a worthy successor to Coutts' efforts.

What Henry Ford did for car manufacturing, our own Morton Coutts did for beer brewing.

For the millions of years before Morton Coutts (b. 1904, d. 2004) was alive, beer was brewed in pretty much the same way. The basic method was developed by the Egyptians, who in turn taught the Romans. They, in the process of conquering the rest of Europe, also brought the gift of beer... and good roads, apparently.

To make beer, barley is soaked in water and allowed to sprout, then the sprouting process is stopped and the result is called malt. This malt is then crushed and mixed with hot water, and the resulting liquid taken away and renamed the wort. Hops are added and the whole mess has yeast added to it. The yeast grows and grows until there is five times as much as there was initially. The fermentation process starts, converting sugars into alcohol. This carries on for a few days, then the beer is aged for weeks or months before being bottled, canned or barrelled, and shipped to pubs, where you buy it, drink a lot of it and wake up with a sore head.

There's nothing wrong with this process, except that it takes a while and in the 1950s and '60s people were drinking more and more beer in New Zealand, and Coutts's employer, Dominion Breweries, was keen to quench their thirst.

Luckily Coutts came at the problem with the ideal family history. Coutts' grandfather was Frederick Joseph Kühtze, a man of German heritage who brewed beer for the goldminers in Otago last century, before moving to Palmerston North to start his own brewery – and why not? The people of Manawatu like a drink as much as the next man. His son, Morton's father, William Joseph Kühtze, inherited the business and promptly shifted it to Taihape, where he correctly surmised that the men working in the timber-milling and bush-clearing operations in the area could do with some amber nectar at the end of a hard day. William also changed the family name from Kühtze to the less-Germanic Coutts – remember that this was around the time of World War I. The name Kühtze lived on, however, as brand name of a lager put out by Dominion Breweries in the 1970s and 80s.

William died in 1918 of the flu, so Morton, aged only 15, took over the business. Talk about a recipe for disaster – would you leave *your* teenager in charge of a brewery?

As it turned out, it was a recipe for something entirely different, for Morton had a brainstorm. If the processes of growing the yeast and fermenting the alcohol could be separated, the whole brewing process could be made continuous. He began with clearer wort – and isn't that what any of us would want? no one likes the idea of cloudy wort – and worked out a way he could keep shoving yeast in

The continuous fermentation system that Coutts invented is now commonly used worldwide, and the system is the mainstay of many Kiwi beers.

one end and have beer come out the other. Through judicious control of the amount of oxygen given to the yeast at different stages, he could encourage it to either grow or ferment. The brewing process was slashed from 15 weeks to 18 hours.

Luck was on the side of continuous fermentation – DB were refurbishing their breweries at around the same time, and when they saw the advantages of the system they converted their production to use this technique. Also fortuitously, and perhaps not too coincidentally, the way the government excise of alcohol worked at this time meant there was less tax on continuously fermented beer.

The final stamp of approval came in 1968, when a beer created with this technique won the 1968 Commonwealth Beers Championship Cup, a much sought-after prize. The continuous fermentation system that Coutts invented is now commonly used worldwide, and the system is the mainstay of many Kiwi beers. Coutts also inspired generations of Kiwis, who see beer as the third R in our national credo. Among them we can count Ian Williams and Anders Warn, who recently invented the world's first personal brewery – the WilliamsWarn. It's a home appliance that allows you to make cold, perfectly carbonated, clear, professional-quality beer, in just seven days.

As an addendum, you may be interested to know that Morton Coutts was also the first person in New Zealand to broadcast television signals, and the first to send a shortwave radio message to Britain. And he also continued to invent new ways to make beer – in 1993 at the age of 90, he registered another patent, improving on the brewing process again. I guess, at least figuratively speaking, beer was in his blood, and it must have been good, because he died in 2004 at the well fermented age of 100.

The Referee's Whistle

WHEN RUGBY MEETS DOG TRIALLING

Kiwis didn't invent the whistle, nor did we invent rugby, or even sport in general, but invention is not always about creating something new; it can be about using existing things in new ways.

In 1884 William Harrington Atack was refereeing a game of rugby in Canterbury using the accepted mode of the time, namely, yelling at the players when they did something wrong. This approach was far from ideal, straining the voice and exhausting the ref. Putting his hand in his pocket he discovered he'd left his dog whistle there, which sparked an idea. The next game he refereed, he sought the players' permission to use it, and the world's first sports game was played 'to the whistle'.

The idea, being so simple yet so clever, was quickly adopted worldwide and soon incorporated into the game's rules – and the rules of almost every team sport played now. It's a shame Atack hadn't bought shares in a whistle factory. On the other hand, it's lucky he didn't have his kid's kazoo in his pocket on that day.

By all accounts, Atack was a born referee – he was described by his family as 'precise and demanding'. He was born in England, and moved to New Zealand at the age of two. He was a sportsman himself, representing Canterbury at cricket and working professionally as a sports journalist. There is a story told of him that once, during a trip to San Francisco, he asked a hotel bellboy to post a letter for him. Unsatisfied with the boy's vacillating over what time the post would be cleared, Atack decided to take the letter to the post office himself. When he returned, he found the hotel had been flattened to the ground – one of the many victims of the 1906 San Francisco earthquake.

The idea, being so simple yet so clever, was quickly adopted worldwide and soon incorporated into the game's rules – and the rules of almost every team sport played now. It's a shame Atack hadn't bought shares in a whistle factory.

Given the ubiquity of the whistle in sport, this may well be one of the most widely used of New Zealand's innovations. Atack's 'invention' in hindsight may seem to be obvious, but don't many of the best ideas?

The Willie Away

NOT AS RUDE AS IT SOUNDS

Right: All Black Wilson Whineray reaches for the ball. New Zealand versus France at Eden Park, Auckland, 22 July 1961. We won 13–6 over the French.

New Zealanders have given hundreds of innovations to the sport of rugby, many intangible (free running back play), others less so (the haka). One is a move called the 'Willie Away'.

In 1961 the French rugby team were touring New Zealand. The tourists came up against the All Blacks three times in tests during that tour. Let me give away the ending here and allay any fears by saying we beat them in every one – but the French certainly gave the new Zealanders a run for their money.

The all black captain at the time was the great player Wilson Whineray (b. 1935, d. 2012), later Sir Wilson. Whineray, a prop, was a born sportsman. As a youth, soccer, swimming and boxing were among the other sports he excelled at – in fact he ended up winning the New Zealand universities heavyweight boxing title. Whineray, then a farmer by day, represented Wairarapa at 17, became an All Black at 21, and his six-year reign as captain started a year later.

Imagine his surprise at Athletic Park, Wellington, in 1961 when the French team he was charged with dispatching pulled ahead 6–5, thanks mainly to a strange move off the back of the lineout that set them up for drop goals. Whineray watched the move with interest, and after the match (which, as mentioned above, they did ultimately win) he set about working on a move of his own using a similar idea.

The move Whineray came up with involves peelingmen off from the back of the lineout, who then drive up midfield, creating a blind side. The second five becomes the first five for a switch of play, and the fullback comes up on the newly created blind side. This effectively commits all the opposition loose forwards and leaves clear space to move the ball around.

The move was very successful, both for Whineray's rep side, Auckland, where it was employed in the defence of the Ranfurly Shield on more than one occasion, and for the All Blacks. The move was dubbed the 'Willie Away', which became the All Black call for the move, and Whineray's place in the All Blacks history was secured.

Whineray followed in the footsteps of other Kiwi innovators and contributors to the game of rugby: Victor 'Old Vic' Cavanagh (b. 1874, d. 1952) was an innovative coach who led Otago University to a decade of legendary success in the 1920s and invented a number of loose scrum techniques; and his son Victor 'Young Vic' Cavanagh (b. 1909, d. 1980) was reputed to have invented rucking.

Whineray himself remained in the ABs until 1965, then set about forging for himself a very distinguished career outside rugby. He became chairman of the Hillary Commission, chairman of Carter Holt Harvey, chairman of the National Bank of New Zealand, a director of numerous companies including Auckland Airport, Nestlé New Zealand and Comalco, and a trustee of many trusts, including the Eden Park Trust and the Halberg Trust. Sir Wilson was knighted in 1998 for his services to sport and business. He died in 2012, and is still remembered as one of the greatest All Black captains.

Tantrix

OVER A MILLION GAMES SOLD

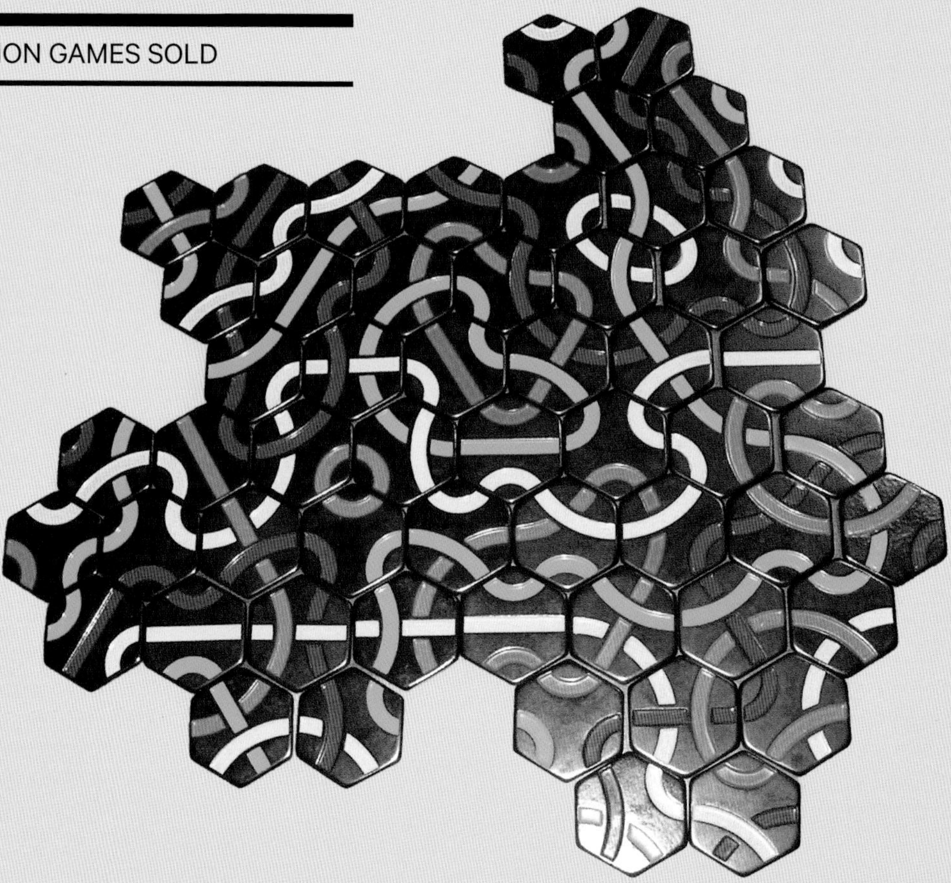

Above: That green loop is winning so far.

To avoid disappointment, let's be clear that this has nothing to do with sex. It is, however, something that consenting adults can enjoy in the privacy of their own home – a game and puzzle called Tantrix created by Mike McManaway. The game comprises 56 hexagonal tiles, which have on them different-coloured lines and curves that must be laid, according to the game's rules, against tiles that match up, making continuous lines and loops. The winner makes the longest single-coloured line or the biggest single-coloured loop (loops are worth more points). Tantrix has an element of chance as well as being strategic, which means that even a beginner or a child will sometimes beat a champion. This luck factor in Tantrix is just enough for it to be a family game. Over a few games, however, the better player will always prevail – and this makes it great for serious competitors. McManaway has dedicated himself to making Tantrix an internationally popular game. He has been not only the inventor, but also the retailer, wholesaler, distributor, player, marketer and evangelist, and he's not interested in inventing anything else. He is not interested in the quick buck, but in putting the money back into the game to ensure it flourishes – and it's working. He's sold a million copies of the game, and created a popular website where thousands play together online.

Tantrix has an element of chance as well as being strategic, which means that even a beginner or a child will sometimes beat a champion. This luck factor in Tantrix is just enough for it to be a family game.

Booktrack

SOUNDTRACKS. FOR BOOKS. YES, REALLY

OK, try using your imagination to not only read, but also hear this bit, and get the full atmosphere of this invention. *Hear* the entrepreneur Mark Cameron exclaiming as he thinks of the idea for book soundtracks while sitting on the Devonport ferry, reading his novel as serendipitously a bright song comes up on his music player. Now *listen* to the tap-tapping of computer keyboards as Mark and his brother Paul spend three years building the idea into the Booktrack, a patented product that will play a soundtrack appropriate for the book you are reading on your device (iPad, Galaxy Note, etc.). You are now *hearing* the soundtrack of the book change as you read, so that the noises (music, atmosphere, sound effects) enhance the passages you are reading. Finally *listen* in to the whoops of joy as the Cameron brothers and their business partners get significant investment for their idea, critical acclaim for the concept, and the beginnings of commercial success.

If you did that right, the soundtrack you created should have significantly added to the reading experience – at least, that's the idea of Booktrack.

Soundtracks for books is a concept that can take some getting used to, but the Booktrack product has been called 'the future of reading' (*The Atlantic*, August 2011). Fans of the product swear by it, saying it adds layers to their reading experience. Investors seem enamoured of the idea, with Booktrack receiving investment capital from some big names in the venture capital world such as PayPal founder Peter Thiel. Salman Rushdie even came to the launch party. Booktrack went on to launch a system called Booktrack Studio to allow authors to create their own musical backings and publish them into a library for purchase and download, shifting control back to the authors and encouraging a new wave of self-published books. As founder Paul Cameron puts in, 'At Booktrack we have been reimagining reading for today's readers, and now with Booktrack Studio we are reinventing writing too.'

As a product, it also has its detractors, with some commentators and press saying the idea 'stinks' and suggesting it's a fad. But then, didn't Gutenberg also get criticised the last time that reading was reimagined? Ultimately, for a product like Booktrack, what the critics say is irrelevant – the product's success will be decided by the market it serves: the authors who publish books with soundtracks, and the consumers who read and listen.

Fans of the product swear by it, saying it adds layers to their reading experience. Investors seems enamoured of the idea, with Booktrack receiving investment capital from some big names in the venture capital world.

The Hotcake

A FAT SOUND THAT RETAINS THE ORIGINAL CHARACTERISTICS

1976

Paul was the electronics whizz of Split Enz, and on tour in 1976, when Noel Crombie decided his guitar sounded too professional, Paul came up with the Hotcake.

Right: Rather disappointingly, the Hotcake looks nothing like a cake.

In New Zealand, Paul Crowther is best known as the drummer for legendary rock band Split Enz. Overseas, Crowther is now best known as the inventor and manufacturer of legendary rock guitar pedal, the Hotcake.

The Hotcake, like other guitar pedals, is a small box that sits at the feet of the guitarist, and can be switched on and off while the musician is in full swing. It adds an effect to the sound of the guitar and makes that distorted, fuzzy sound (mimicking the sound of an overloaded amplifier, without risking damage to the amplifier) that has been at the heart of rock and roll forever. The idea is to spend thousands of dollars amplifying the sound of a guitar perfectly and then ruin it. But ruin it just right. The effect produced depends on the electronics inside the box, and the Hotcake is globally renowned for its brilliant sound. In music-speak, most distortion effects provide distortion at the expense of the clean, original character of the guitar. The Hotcake retains the underlying quality of the guitar while providing a nice, fat, thick distortion that guitarists go crazy over.

Crowther was the electronics whiz of Split Enz, and on tour in 1976, when Noel Crombie decided his guitar sounded too professional, Crowther came up with the Hotcake. It turned out to be something brand new, and other musicians began to demand them. Crowther started making more and more, and began distributing them to New Zealand music shops in 1977. When international bands came to New Zealand, Crowther would often do the sound engineering at their gigs and often they would be so amazed with the sound, they would take a Hotcake home with them. In 1994, overseas music guru Ken Fischer made a mention of it in his *Vintage Guitar* magazine column and international mail order sales began to climb.

The list of the bands that use the Hotcake pedal includes almost any New Zealand act ever – Split Enz, Dave Dobbyn, the Chills, Crowded House, Chris Knox – and any overseas act worth their rock'n'roll salt too including Mark Knopfler, Beck, Sonic Youth, Blur, Pavement, Oasis and Radiohead. Crowther has even followed up the Hotcake with another pedal, a 'harmonic generator intermodulator' he calls Prunes and Custard.

At US$180 each, Crowther has sold more than 10,000 hotcakes, and despite sales topping a thousand a year, Crowther and his wife Jo still make the pedals, cottage industry-style, in their home in Auckland. Crowther was certainly very canny – he realised that if you invent Hotcakes, there's only one way they can sell…

Fastmount

SUPER FITTINGS FOR SUPERYACHTS

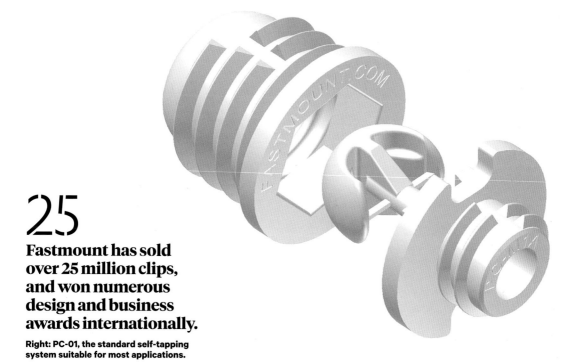

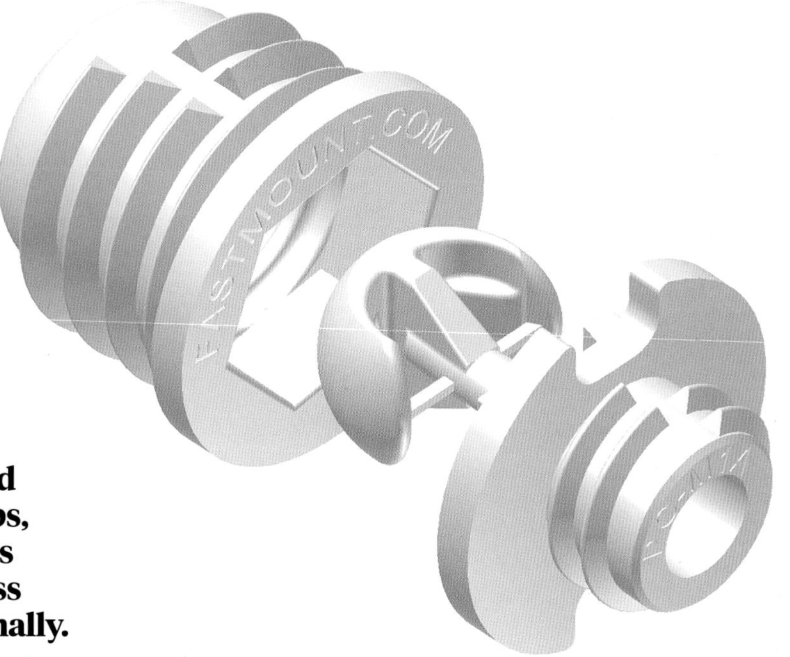

25

Fastmount has sold over 25 million clips, and won numerous design and business awards internationally.

Right: PC-01, the standard self-tapping system suitable for most applications.

New Zealand's high-tech marine industry gave birth (and berth!) to a company that has developed a superbly simple and beautifully designed system for mounting removable surface panels. As you probably know from your own one, superyachts are lined with removable panels so that their internal mechanisms such as electrics can be easily accessed. But Gregg Kelly was frustrated with the normal (often just Velcro) system for mounting interior panels on the yachts. The panels on these multi-million-dollar boats simply never aligned properly and kept falling off. In 2004 Kelly and designer Ron Hanley set themselves the challenge of solving these problems. The result was Fastmount – not just a bunch of fixings but the world's first full mix-n-match system for fitting panels. They've consistently used customer feedback, making their system even more versatile. Says marketing manager Joss Hong, 'Constantly asking "What other panel mounting problems could we solve?" is a key principle to our success.' This 10-year-old business, with just four employees and no office – they work wherever there is Wi-Fi – has survived the global financial crisis and is gearing up for growth, also expanding to the campervan, aviation, industrial and architectural markets. Luxury yacht industry leader Sunseeker was one of their first international customers and business has grown fast with 1000 boatyards in 50 countries using this revolutionary, made-in-New Zealand system to mount panels. Fastmount has sold over 25 million clips, and won numerous design and business awards internationally.

Live 3D Sports Animations

ANIMATION RESEARCH LTD

The 1992 America's Cup was memorable for New Zealand in more than one way. On the water Team New Zealand only narrowly lost the Louis Vuitton Cup, but on television was where Kiwis really saw their team deliver. The graphics invented by Ian Taylor and his company Animation Research Limited (ARL) for that regatta were ground-breaking and are still leading the world in real-time sports animation today.

To be fair, if there was ever a sport that needed to tweak its television appeal it was yachting. The sport had all the visual excitement of cauliflower. Then ARL's graphics came along and allowed viewers to see the context of the race – zooming out to show the whole course, with the yachts' paths superimposed with lines to show which yacht was leading.

The self-effacing Ian Taylor will tell you the success they had was all about assembling the right team –

Below and opposite: Animation Research made yachting understandable for all.

500,000

Taylor took the idea to TVNZ who borrowed a $500,000 computer and ARL created their Winged Keel software for the 1992 America's Cup. In doing so they invented the world's first live animated graphics for sports coverage.

a team who, whenever challenged by being asked if they could do something, always answered, 'I don't see why not.'

ARL started in 1989 as a joint venture between Taylor and the University of Otago, who provided him with a team of international prize-winning computer programmers. They made the opening graphics for *University Challenge* and then went on to graduate to other challenges.

Just sitting around in the office, Ian posed the question to his team: 'Is there anything we can to do make watching yacht racing easier?' One of his programmers said, 'If there's data coming off those boats I don't see why not.' Taylor took the idea to TVNZ who borrowed a $500,000 computer from Silicon Graphics for the job and ARL created their Winged Keel software for the 1992 America's Cup. In doing so they invented the world's first live animated graphics for sports coverage. It changed the way the world watches sports on TV. Taylor plays it down. 'Our big innovation? Showing who was leading.'

Since then they have continued to make breakthroughs – the first live-to-air GPS ball tracking for golf, live animated graphics for golf and cricket, the skiing Bluebird penguins, and Virtual Spectator, which since the 2000 America's Cup has delivered an animated version of yacht racing live to the world via the internet.

Taylor is a big fan of New Zealand's No. 8 Wire attitude. But while the internet means New Zealand is no longer so isolated, we still need to encourage our special way of thinking. 'It's not about the wire,' he says, but about using unconventional tools to solve problems in unexpected ways, and about doing things you're not supposed to be able to do. 'F... the boxing, just pour the concrete!'

If any criticism could be levelled at ARL it's probably that the graphics are a bit too realistic. New Zealanders would no doubt appreciate a version of the software where, instead of being always totally 'true to reality', Team New Zealand wins.

The Shweeb and the Zorb

WORLD'S FIRST HUMAN-POWERED MONORAIL AND A PLASTIC HAMSTER-BALL FOR HUMANS – WHY NOT?

89

As part of the Rotorua Agroventures Adventure park, their fully working concept model Shweeb has had over 75,000 riders from the ages of four to 89, and shows the viability of the concept in a fun way.

Left: The Zorb, turning excess money into thrills and spills.

Imagine riding a bike, lying on your back, encased in plastic, suspended four metres above the ground. Now remove the handlebars, wheels and ability to go wherever you want – and you have the Shweeb! Contrary to what you are thinking, the word 'shweeb' is a derivation from the German verb 'schweben', meaning 'to float, or suspend' – it does *not* come from the English word 'dweeb', which means to lie flat on your back trying to pedal in a plastic bubble.

But apparently the wise heads at Google were somewhat less cynical than you when it comes to the Shweeb – they invested $1m into the idea after the Shweeb was entered into their Project 10^100. This project sought new ideas that would help the world, and the five winners included the Rotorua-

Above: People shweebing their hearts out around a corner.

They didn't invent the idea of putting people into a giant ball and rolling it down a hill, but they found a way to make people pay for it. Their venture began in 1994 in Rotorua and people now Zorb all over the globe.

based invention. The Shweeb in particular was picked for its potential use in future urban environments, as a clean green monorail system for short-to medium-distance personal transport.

Geoff Barnett conceived of the system in the urban jungle that is Tokyo in 2008. Originally from Australia, he relocated to the Rotorua lakeside to work with his Kiwi business partner Peter Cossey to turn Shweeb into a reality. As part of the Rotorua Agroventures Adventure park, their fully working concept model Shweeb has had over 75,000 riders from the ages of four to 89, and shows the viability of the concept in a fun way.

It's near that other Kiwi innovation, the Zorb – a device designed to induce vomiting in adults with too much money. Brothers David and Andrew Akers didn't invent the idea of putting people into a giant ball and rolling it down a hill, but they found a way to make people pay for it. Their venture began in 1994 in Rotorua and people now Zorb all over the globe.

Working with Google, and with engineers around the world, Cossey and Barnett have continued to refine both the Shweeb and their overall business model. Cossey describes the process of creating a new urban transport system as very time-consuming and expensive, with lots of learning along the way and some negativity to overcome. He prefers to take the approach of assuming the idea *will* succeed, and then working around the reasons why it wouldn't. The Google money helped, but he says it's just 10 per cent of the money they need to realise their vision.

They now have partnerships in Europe, the United States and Korea to develop the Shweeb both as an adventure ride, and as an urban transport option which they hope will one day give the same flexibility and comfort offered by the car, with fitter, healthier passengers and no environmental impact. No dweebs allowed.

The Tote

ODDS ON HE'S A KIWI

Oh, boy. This is a good one! An Australian website loudly proclaims 'Totalisator History: An Australian Achievement', but in fact, the invention of the automatic totalisator is as much a New Zealand one as it is Australian. It is a story of the birth of the computer, of one of the most lasting social debates in New Zealand history, and of the relentless quest to separate people from their money faster.

Gambling has always been a big part of New Zealand life, and horse racing has long been our favourite way to lose money. New Zealanders love

Mr Julius converted the idea from one that serves democracy to one that fleeces people out of their money. Despite never betting on horses, he asked a friend to explain the totalisation process to him and then he invented the automatic totalisator.

1913

In 1913 the first automatic totalisator in the world was installed at Ellerslie Racecourse.

Left: Punters lining up to place a bet on the automatic tote, Auckland, 1952.

to try and pick which horse will be able to run the fastest. In 1913 the first automatic totalisator in the world was installed at Ellerslie Racecourse. When punters place a bet on a horse, the amount of the bet must be recorded. The amount each horse has placed on it is tallied and at the end of the betting, the odds are calculated by what proportion of bets are placed on each horse. This process is called totalisation. Before 1913, these calculations were carried out manually. It was a slow and inaccurate process, and the race could not begin until all the bets were totalled and the odds calculated. Bookies could run a service (from their spots standing on wooden boxes) just as smoothly. So, despite being legally banned from racecourses in 1911, the bookmakers still pulled in money – probably just as much as the racecourse totalisators.

George Julius (b. 1873, d. 1946) was born in England, went to high school in Australia, then came with his family to Christchurch where his father was the Archbishop of New Zealand. He attended the University of New Zealand in Christchurch and in 1896, with his engineering degree, he moved back to Australia. He worked as an engineer, but in his spare time (as many of us do) he came up with a machine to automatically count electoral votes. This machine was turned down by the Australian government. It was ahead of its time – in fact way ahead, in fact it may never have a time – votes in New Zealand to this day are still mainly lodged and counted manually. Mr Julius (later to be knighted for his services to technology) converted the idea from one that serves democracy to one that fleeces people out of their money. Despite never betting on horses, he asked a friend to explain the totalisation process to him and then he invented the automatic totalisator.

In 1908 Julius offered his invention to the Ellerslie Racecourse, and in 1913 the Auckland Racing Club held the world's first race meeting where the bets were automatically recorded and computed. It was a huge machine, filling its purpose-made building with what looked like a giant tangle of piano wires, pulleys, sprockets, cast-iron boxes, and heavy weights hanging on wires for drive power. It had 30 booths, at any of which a punter could place a bet on any horse. The booths printed a ticket out for each punter and were connected mechanically by wires to the totalisator machine, which kept a mechanical tally. The tallies were displayed on the front of the tote building for all to see. When betting was stopped the totals and odds were already calculated. Brilliant! It was effectively a huge computer with 30 terminals and one big display.

Despite a lot of scepticism, it was a huge success. Julius formed a company and had a world monopoly for many years, installing hundreds of his machines all around the planet. By 1970, almost every major racing centre in the world used Julius's Automatic Totalisators, which were in service in 29 countries. At first they were purely mechanical, then mechanical and electrical; they finally became computers in the 1960s. They also began to put paid to the bookmaker's trade, which disappeared after off-course betting was made legal and controlled by the introduction of the TAB.

Is it fair to call the automatic tote a New Zealand invention? George Julius was born in neither New Zealand nor Australia, but England. His degree was from Canterbury and his machine was first installed in Auckland. So despite the fact that the machine was invented and manufactured in Australia, I think we can safely claim it. We'll leave the tote machine with a quote from Sir George Julius himself. See if you think he sounds Kiwi or Aussie: 'Up to that time I had never seen a racecourse. A friend... explained to me what was required in an efficient totalisator. I found the problem of great interest as the perfect tote must have a mechanism capable of adding the records from a number of operators all of whom might issue a ticket on the same horse at the same instant.' He's a Kiwi.

9

Weird Science

ELECTRIC PLASTIC, BLACK HOLES
AND THE BRAIN HIGHWAY

Why are our internationally famous scientists obscure in their own country? Maybe it's simply because there are no famous brands of scientists, science is not an Olympic event, and science isn't on the telly. It's more than a curiosity – this absence of knowledge and pride in science is actually a serious inhibitor for our growth. Our future lies in applying science, engineering, technology and mathematics to problems of today and the pressing needs of the future.

The No. 8 Wire that future inventors will manipulate into the inventions of tomorrow will be things like conducting polymers, carbon nanotubes, magnetic fields and superconductors; and the farmers bending them to unexpected uses will be our scientists.

The (ironic!) title of this chapter perhaps highlights the problem – science, and scientists, are 'weird' to most of us. They talk in language we don't understand, they focus on the very small or the very large and it's hard to see what they get excited about or how it affects our everyday world. But our scientists are, and have always been, among the best in the world in their fields, and science, simply, changes everything.

The Kerr Metric

THE MATHEMATICS OF ROTATING BLACK HOLES

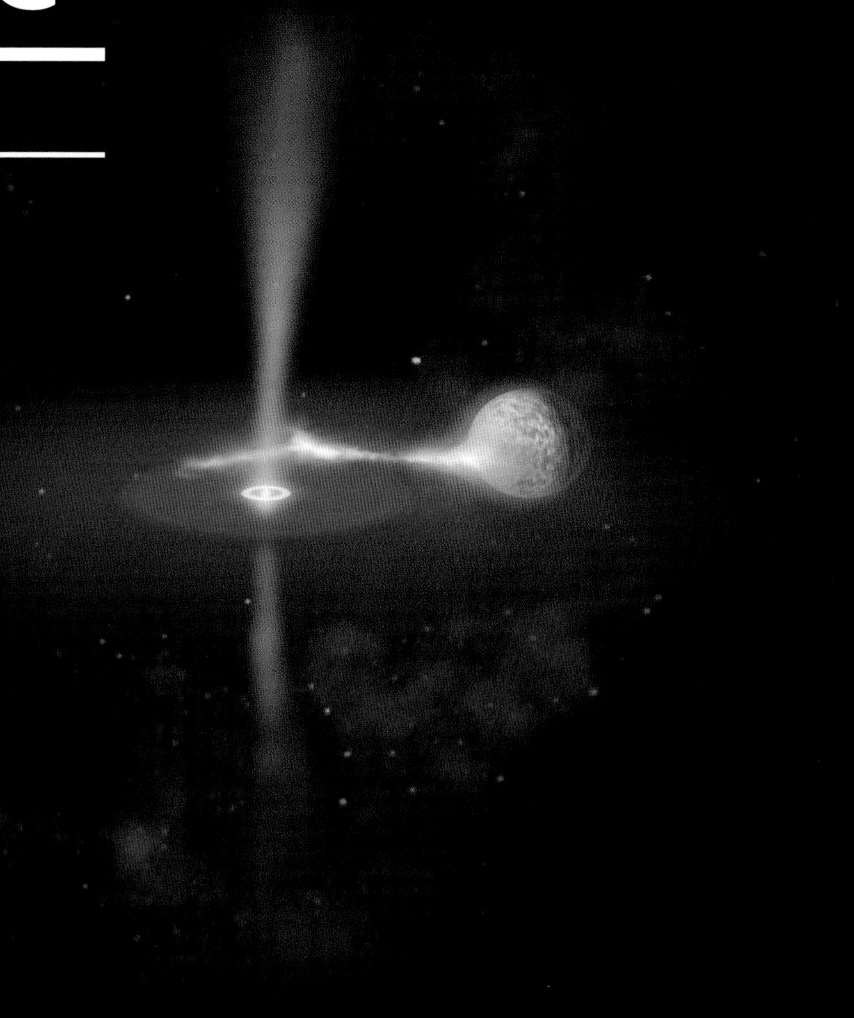

40

The sheer difficulty of the maths involved meant that for 40 years after he wrote it out, no one had solved the equations at the heart of Einstein's theory.

Right: Kerr's work looks at the weird and wonderful areas around black holes.

In 1930 Albert Einstein, a patent clerk in Switzerland, started work on his Theory of General Relativity. In 1963 Roy Kerr (b. 1934), a 29-year-old maths graduate from New Zealand, finished it.

Well, kind of.

When Einstein penned his revolutionary theory, he left a number of 'holes'. It was later discovered that these holes were quite literally holes – stellar phenomena that you have probably heard referred to as black holes. These were initially thought so weird and exotic that they couldn't exist in the real world, but over time scientists came to realise that not only did they exist in the cosmos, they are also quite common, with one even at the heart of our own galaxy.

Black holes are basically very heavy things – heavy enough that their gravitational pull is so strong that everything that comes close to them, even light, gets sucked in and can't escape. They eat up stuff around them, and given that they are so incredibly heavy – a teaspoon of the stuff would be many times heavier than the whole Earth – they distort space and time in the surrounding area. It's like

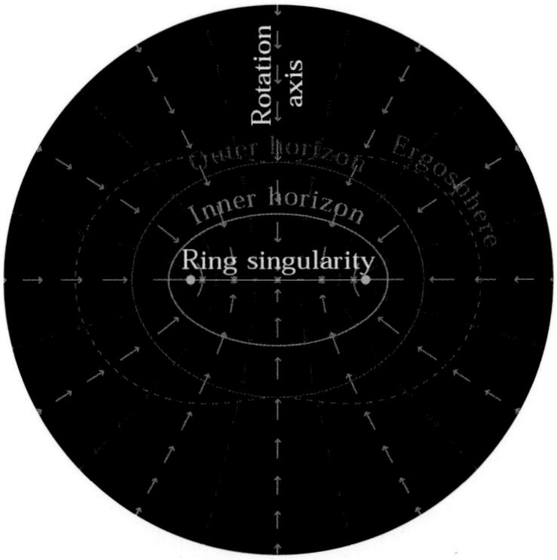

having really bad neighbours who bring down the entire neighbourhood. It is in the study of this 'neighbourhood' that Professor Kerr made his contribution. In fact, to be entirely accurate, it's the study of rotating black holes, so it's kind of like neighbours who bring down the neighbourhood and at the same time spin at incredible speeds. Maybe this whole neighbourhood metaphor is kind of falling apart – but then so does the fabric of space around a black hole, so it's kind of appropriate.

The sheer difficulty of the maths involved meant that for 40 years after he wrote it out, no one had solved the equations at the heart of Einstein's theory. The six interlocking equations were seen as the most difficult to crack of all time. What Kerr did was to sit down and write some very complex equations which model very accurately what goes on near a black hole. Scientists use Kerr's equations to guess what black holes might do as they pass through space, and science-fiction writers ignore Kerr's equations when they make spaceships fly into them and arrive in another dimension. If Arthur C. Clarke had just done his maths, that whole *2001: A Space Odyssey* thing would never have happened.

From the time of Kerr's announcement of his theory, it proved to be an enormously useful way of describing what scientists call 'spacetime'. The Kerr Metric (also known as the Kerr vacuum solution) is still the method used to examine the complex geometry of the region surrounding black holes.

Kerr was born in Gore, and spent much of his career from 1955 onwards studying, then working, in universities overseas. He did his work on the solutions to general relativity while in Texas. He returned to New Zealand in 1971 and worked as professor of mathematics at the University of Canterbury until his retirement in 1993. In 2012 Kerr became the first New Zealander to be awarded the Einstein Medal, presented annually to individuals for outstanding service, discoveries or publications related to Albert Einstein. Kerr is also a winner of the Rutherford Medal, and is a Companion of the New Zealand Order of Merit for services to astrophysics.

For completeness, and to prove we've done our research, we've included the equation for the Kerr Metric (written in Boyer-Lindquist coordinates of course). What does it all mean? We have no idea. We'd venture to suggest neither do you, but we don't think this makes Kerr's innovation any the less important or noteworthy.

THE KERR METRIC

$$ds^2 = -a^2 dt^2 + \varpi^2 (d\phi - \omega dt)^2 + (\rho^2 / \Delta) dr^2 + \rho^2 d\theta^2$$

where the coordinate functions are given (with G=c=1):

$$\Delta = r \quad + a^2 - 2Mr$$

$$\sum^2 = (r^2 + a^2)^2 - a^2 \Delta \sin^2 \theta$$

$$\rho^2 = r^2 + a^2 \cos^2 \theta$$

$$\omega = \frac{\sum}{\rho} \sin \theta$$

the specific angular momentum is:

$$a = \frac{GJ}{Mc^3}, 0 \le a \le 1$$

Incidentally, the physical value of J is for a star like our sun:

$$J = 1.63x10^{48}\, gcm^2 / s$$

Corresponding to a=0.185 M. If a=0 we have the Schwarzschild case for a non-rotating black hole (or star). So there.

WASSP

FLASH FISH FINDING

120

Their patented transmitter sends pulses out eight times per second, and covers the seafloor at a 120 degree angle giving amazing 3D images of what's going on down there.

Below: The WASSP output showing depth and terrain – and fish!

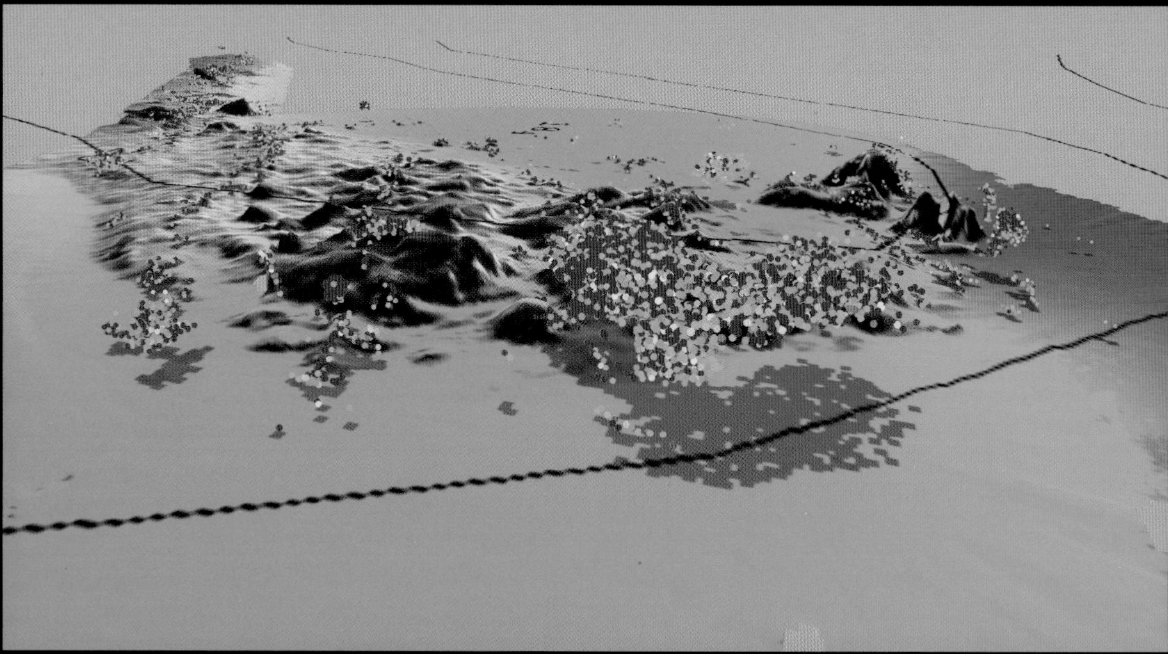

This collaboration between Callaghan Innovation (ex-IRL) scientists and Auckland company Electronic Navigation Limited (ENL) took a while. More than 15 years of research, development and testing went into the Wide Angled Sonar Seafloor Profiler or WASSP. Their innovations in signal processing, and especially in the transducer which goes under the boat – they call it the 'wet end of things' – have resulted in a product that is vastly better than your average fish finder. ENL wanted a product that was better than the weekend tinny drivers use, but cheaper than military gear. They wanted to profile the seafloor with a single beam, and show results in glorious colour 3D. Their patented transmitter sends pulses out eight times per second, and covers the seafloor at a 120 degree angle giving amazing 3D images of what's going on down there. It's used now by the super-yacht industry as well as by commercial fishermen, survey craft and other workboats and it's selling in over 40 countries around the world.

KiwiStar

SHONE BRIGHT FOR A SHORT WHILE

Right: The colourful KiwiStar lens.

Like many of us in 1986, David Beach stood outside his home at night in the cold, looking skyward for a glimpse of Halley's Comet. And like many of us, he was disappointed in this once in a lifetime event – imagination played quite a large part in the proceedings. Beach tried photographing the comet, but again with disappointing results. The comet was too far away, and the light from it was too dim. But unlike the rest of us, Dr David Beach was a physicist working in Auckland for Industrial Research Ltd's imaging and sensing division, with training and experience in refraction techniques and optics. He set about creating what was to become an incredibly powerful lens system.

The KiwiStar system was a new type of ultra-fast camera lens with high resolution, broad spectral bandpass and high scalability (100mm to greater than two metres), which has no low-order aberrations. In terms you and I can understand, it's a very high quality lens that can take crystal clear pictures over a long distance in low light – the best ever telescopic camera lens. KiwiStar had applications in many areas, from astronomy to surveillance – the KiwiStar lens could photograph a car number plate from a kilometre away at night, making it an ideal tool for police and the military. The nuclear facility at Los Alamos in the USA immediately bought one to make images of the core of a nuclear weapon.

It was to be the only KiwiStar sold, however, because in the late 1990s, shortly after Beach perfected KiwiStar, advances in computer-controlled glassmaking started a revolution in lens manufacture. Aspherical lenses, always the best for extremely sharp images, had previously been too expensive to use in any sensible system, but new ways of making them brought the price down and KiwiStar's window of utility closed tight. At the time, the lens made such a splash worldwide that IRL (now Callaghan Innovation) named their optics division KiwiStar, and though the KiwiStar lens itself was shelved, the division continues to win multi-million dollar contracts making major optical components for important telescopes around the world. So maybe, in 2056, when Halley's Comet arcs its way back into view, we'll all be able to get some decent photos, thanks in part to Dr Beach.

In terms you and I can understand, it's a very high quality lens that can take crystal clear pictures over a long distance in low light – the best ever telescopic camera lens.

The Topology of Knots

DONUTS AND COFFEE CUPS GIVE A KIWI THE FIELDS MEDAL

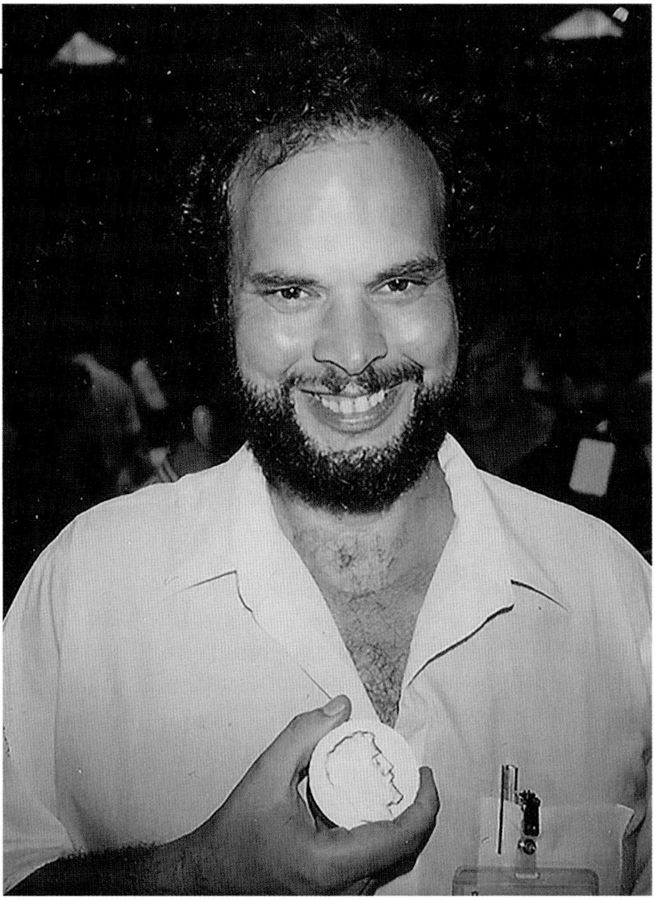

1990

When Jones was awarded his medal in 1990, it was specifically for the discovery of a new relationship within the realm he was studying: a polynomial invariant that apparently had been missed by many others who were studying in the same area.

Right: Jones with his Fields Medal in 1990.

The Fields Medal is often referred to as the Nobel Prize of mathematics. It is awarded only once every four years to up to four mathematicians worldwide for significant contributions to the study of mathematics by those under the age of 40. The only New Zealander ever to receive one is Sir Vaughan Jones (b. 1952), who received his Fields Medal in 1990 for work done in the mathematical area called topology.

Topology is a branch of mathematics regarding shapes. It is often called rubber sheet geometry. It's the idea that in topology shapes are defined not by their geometric shapes, but by their complexity. Thus a circle and a triangle in 2-space are topologically identical, both containing one hole and so being equally complex; so too are a donut and a coffee cup in 3-space – maths language for the real world. Don't worry, it's pretty esoteric stuff, and like a lot of pure mathematics, actually more resembles gibberish than it does plain English ('Let M denote a von Neumann algebra. Then M is an algebra of bound operators acting on a Hilbert space H…')

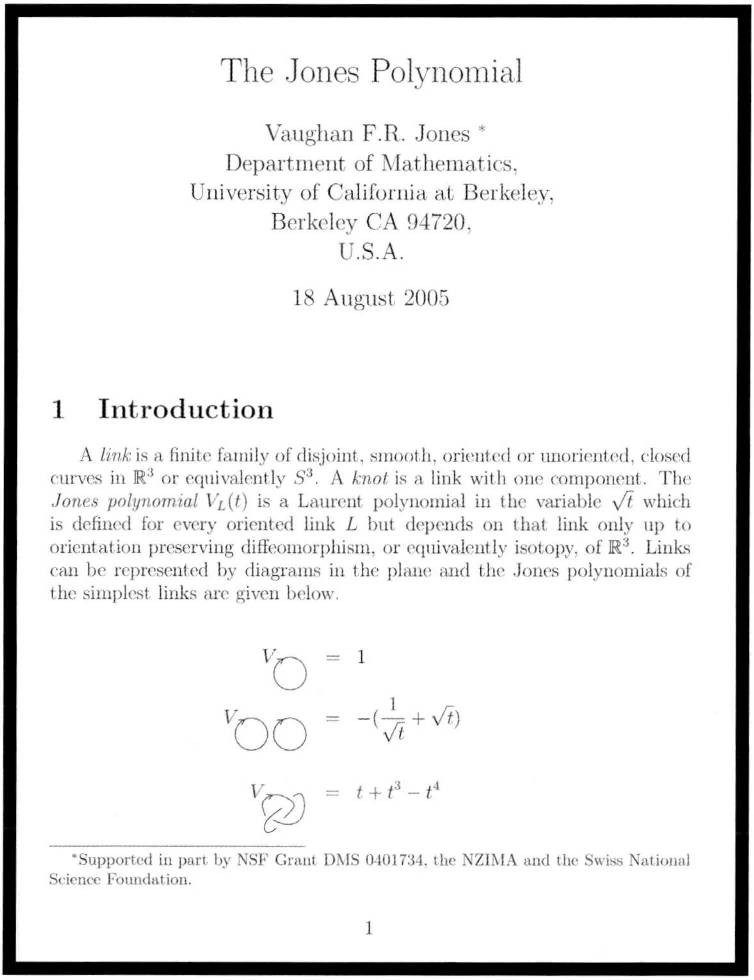

The Jones Polynomial

Vaughan F.R. Jones *
Department of Mathematics,
University of California at Berkeley,
Berkeley CA 94720,
U.S.A.

18 August 2005

1 Introduction

A *link* is a finite family of disjoint, smooth, oriented or unoriented, closed curves in $\mathbb{R}^3$ or equivalently S^3. A *knot* is a link with one component. The *Jones polynomial* $V_L(t)$ is a Laurent polynomial in the variable $\sqrt{t}$ which is defined for every oriented link L but depends on that link only up to orientation preserving diffeomorphism, or equivalently isotopy, of $\mathbb{R}^3$. Links can be represented by diagrams in the plane and the Jones polynomials of the simplest links are given below.

$V = 1$

$V = -(\frac{1}{\sqrt{t}} + \sqrt{t})$

$V = t + t^3 - t^4$

*Supported in part by NSF Grant DMS 0401734, the NZIMA and the Swiss National Science Foundation.

1

Above: The Jones Polynomial is now a key paper in mathematics literature.

Jones' Fields Medal recognised him as an inventor of mathematical machines, an engineer making things out of numbers.

Jones was born in Gisborne, but like many others at the top of their field, left New Zealand after his initial studies and went to work and study overseas, in a variety of countries and universities. Always, though, it seems Jones typified the New Zealand spirit – he has been described as 'informal' and 'encouraging the free and open exchange of ideas'. In the admirable tradition of mathematicians, he often gives out details of what he is working on to allow others to contribute their thoughts, not jealously guarding his own ideas to hog all the glory.

When Jones was awarded his medal in 1990, it was specifically for the discovery of a new relationship within the realm he was studying: a polynomial invariant that apparently had been missed by many others who were studying in the same area. This is akin to a whole lot of gold prospectors working the same stretch of river when one of them strikes gold. Although in this case the gold was due to hard work and inspiration as much as luck. Jones' Fields Medal recognised him as an inventor of mathematical machines, an engineer making things out of numbers. As a nice finale, apparently Jones caused quite a stir in 1990 when he attended the International Mathematics Congress to pick up his prize, giving his valedictory lecture dressed in the All Black rugby strip. I wonder what number he had on his back. Probably not 2 or 15; more likely $\sqrt{\frac{i^2}{2}}$.

And apparently the All Blacks often reciprocate by holding coffee cups and eating donuts.

Punchcards for Scientific Analysis

SAVING MILLIONS OF LABORIOUS CALCULATIONS

Right: Comrie lost his left leg while fighting in World War I. While recovering he developed an interest in mechanical computation and its use in astronomy.

I bet you don't have an asteroid and a hole in the moon named after you, but Dr Leslie J. Comrie (b. 1893, d. 1950) does. Pukekohe-born Comrie studied first at Auckland University College, then was sent to fight in World War I where he lost his left leg to 'friendly fire'. In 1923 he received his PhD from Cambridge, and in 1926 he became the deputy superintendent of His Majesty's Nautical Almanac Office in Greenwich, London. Part of the function of the Greenwich institution was to calculate the phases and locations of the moon, the stars and other planets; at the time those calculations were done laboriously by hand. Comrie was the first to use the new punched-card equipment for scientific purposes. Punched cards had been developed to control industrial machines like looms, but Comrie realised they could be used to race through the millions of calculations needed to compute astronomical tables. His new techniques allowed these to be done automatically, at a rate of 20 or

Punched cards had been developed to control industrial machines like looms. Comrie realised they could be used to compute astronomical tables, which require millions of calculations.

30 per second. Comrie collaborated with peers in the US and after World War II was influential in applying computers to science and mathematics.

Comrie was quite deaf, and was described as quiet until things got going, and 'then his enthusiasms took over. Can you imagine being excited over which numerical type fonts worked better in a math table?' Comrie's contribution to our precise understanding of the position of the moon led to his name being given to a crater there – at 23·3N, 112·7W, if you're ever passing.

Good Laboratory Practice

QUALITY ASSURANCE IN LAB TESTING

1972

The 1972 New Zealand Act was the first time in the world the term 'good laboratory practice' had been used officially, but over the coming years the idea would spread internationally.

During the 1960s and early '70s the scientific world began to pump out new chemicals and pharmaceuticals at an increasing pace. All over the world labs were set up to test these chemicals and drugs, to make sure they worked and were safe for people to use. There was big money involved in these drugs and a scandal erupted in the United States in 1976 when the FDA found many labs had poor equipment and practices, giving untrustworthy results. Some had actually falsified data (making dangerous drugs appear safe) and many were treating lab animals poorly. The public lost confidence in the labs, people went to jail, and something needed to be done. That's when they turned to a New Zealand idea.

In 1972 the New Zealand government had introduced the New Zealand Testing Laboratory Act to promote, develop and maintain standards in testing laboratories across the country. They called this 'good laboratory practice' (GLP). GLP lays down rules and standards for everything in a lab from the qualifications of the personnel to the hygiene, record-keeping, operating procedures and equipment maintenance. Having a standard means a lab can be audited simply. If it passes, then the lab, and its test results, can be trusted.

The 1972 New Zealand Act was the first time in the world the term 'good laboratory practice' had been used officially, but over the coming years the idea would spread internationally. In 1979 the FDA enforced its own GLP regulations and in 1981 the OECD issued GLP principles for the world. Having new drugs and chemicals to make people healthier and life easier is priceless; and GLP means we can trust them – thanks to us.

Gasified Waste Carbon

THE RED HOT GREENY

85

The world uses around 85 million barrels of oil a day for energy, and it seems like the best idea would be to deal with the efficiency of the energy that is already being used.

Below: The Lanzatech system at work in a plant in China.
Right: Co-founder, Dr Sean Simpson.

Looking at Dr Sean Simpson, you would be forgiven for thinking you are seeing a 'screaming greeny' (his words), with his long red hair and evangelical demeanour – he even bikes to work. But this is a guy who also raises huge amounts of venture capital, practises genetic modification and – worst of all – exploits millions along the way. 'I don't go around hugging trees all day.'

But Simpson, a UK import who now calls New Zealand home, is passionate about finding a commercial solution to so-called 'green issues' – among them finding an economic incentive for industry to solve the problem of waste products. How is it that we just throw away the so-called waste products from so many processes? Not only is that a financial waste, it is also clearly an ecological issue. Lots of work around the world is being done on the idea of sequestration – almost literally digging holes in the ground to put waste gas and chemicals into, with all the attendant risks of leakage and environmental impact. It's silly when you think about it, and clearly something needs to be done. So in 2005 Simpson decided to use his training and experience in bioproduct development to meet the challenge.

Lanzatech was born.

Lanzatech's process takes the waste gas from manufacturing, and feeds it into large tanks of microbes suspended in liquid.

So, too, were the billions of tiny microbes which are Simpson's surrogate employees – in fact the majority of his workforce is comprised of bacteria. In many ways they are the perfect workers – they work 24/7, they never complain, and they take whatever rubbish gets thrown at them and make something of it. In fact they make ethanol – and these little microbes are the secret ingredient in Lanzatech's plan to solve the issue of waste and energy usage.

Simpson's insight was that the issue of waste gas is always going to be here. With the world population growing towards 11 billion in the next 20 years, the use of alternative energy sources isn't really going to be enough. These people will need a lot of food, and the corn, palm oil and other sugar-based energy alternatives are not keeping up with demand. The world uses around 85 million barrels of oil a day for energy, and it seems like the best idea would be to deal with the efficiency of the energy that is already being used. An economically sustainable approach is what's needed.

Lanzatech's process takes the waste gas from manufacturing and feeds it into large tanks of microbes suspended in liquid. Simpson likens it to putting a brewery next to a steel mill. The bacteria consume the carbon dioxide (CO_2) and carbon monoxide (CO) from the waste gas, add in hydrogen (H) from the liquid, and turn it into ethanol (CH_3CH_2OH). The other product of the process happens to be mostly common water, which can be fed back into the tank. The ethanol can then be used in a number of ways – most commonly it will be sold to petroleum companies to be mixed in with other fossil fuels. Brilliant.

The chemistry behind this looks simple, but of course the real smarts is in making the whole process robust, finding the right microbes and proving it will work on an industrial scale. Simpson and his co-founder Dr Richard Forster (b. 1949, d. 2014) set up a research and demonstration version of their idea at Glenbrook steel mill near Auckland. They successfully showed that their microbe process was better for the mill owners than the alternative – burning the waste products to generate electricity. But that proof took a lot of money – at least US$90m has been spent so far to create and refine the technology, and that sort of money can't be found in New Zealand alone. It's a scary amount, but it came in a bit at a time, as they proved their ideas. Simpson remembers getting the first significant amount – $3.5m – and thinking, 'What would happen if we stuffed it all up?' The Warehouse founder Sir Stephen Tindall was one of the first investors to help Lanzatech get off the ground, but the serious money has come from the United States.

While Simpson says they are bizarrely called a 'foreign-owned' company, this isn't a story of a New Zealand company leaving our shores. The company has grown to employ people as well as microbes – over 140 of them, of whom 80 are in New Zealand. Simpson is at times frustrated with the small thinking of the New Zealand psyche – the 'farmer mentality', he calls it. It's served us well, but we need to move on. Simpson sees himself as not just a scientist inventor, but also an entrepreneur, running a technology start-up. His energy and enthusiasm for what he is doing is infectious, and Simpson regularly presents on science and innovation, and on the commercialisation of ideas. He may look like the archetypal mad scientist, but this particular madness is one we can all benefit from. Lanzatech has filed over 200 patents and set up joint ventures and operations in China (with China's second-largest steel mill), Korea, Malaysia, and the US. They've won awards for sustainability, appeared in *Forbes* magazine, met with Barack Obama, and Simpson himself has been recognised as an innovator and champion of sustainability.

Not bad for someone who doesn't pay his workers.

KODE

VIRUS PAINT BY NUMBERS

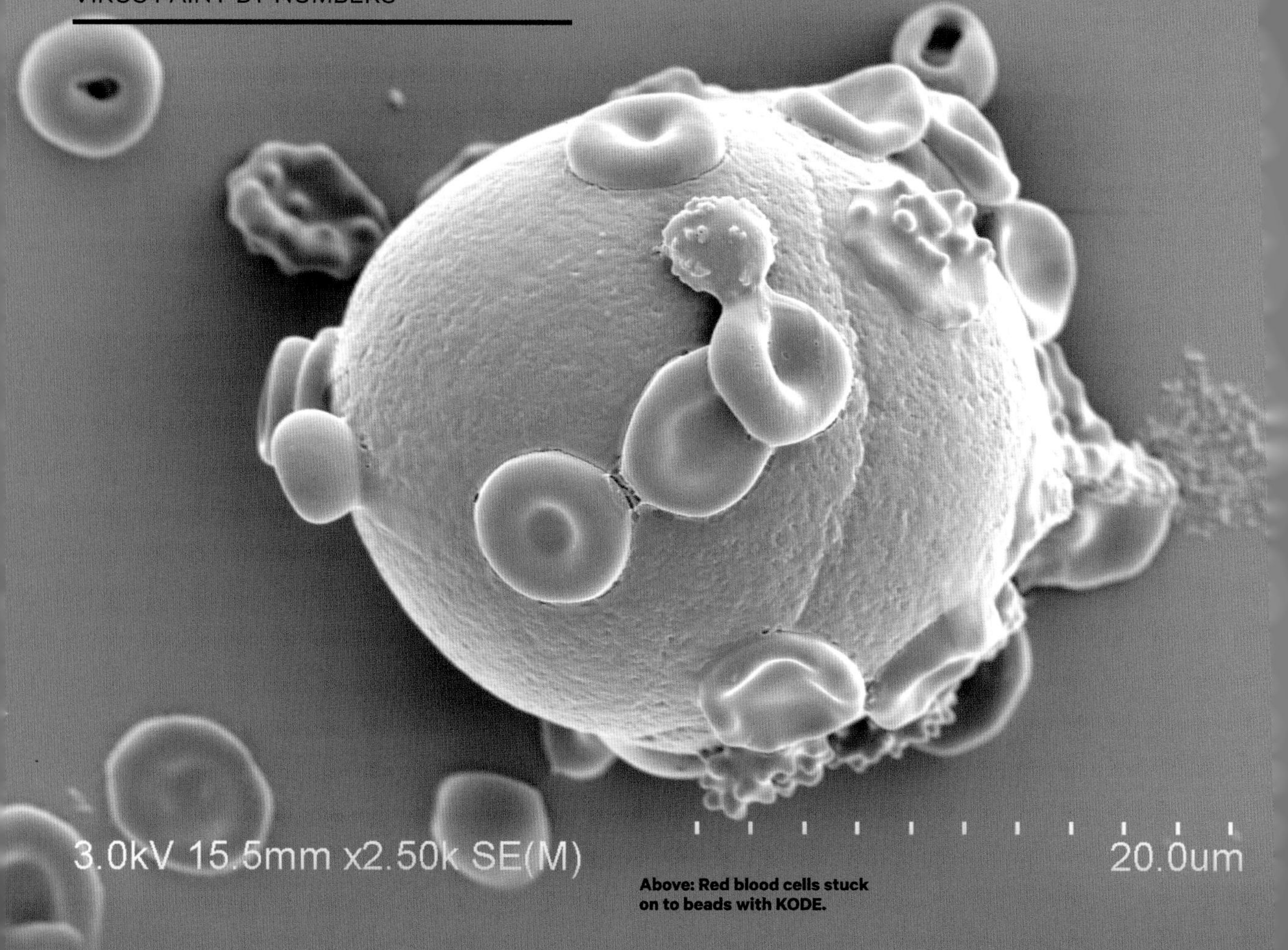

Above: Red blood cells stuck on to beads with KODE.

Professor Stephen Henry is a world champion archer. In 2006 he shared the world record with a perfect score (560/560) in the animal round of the International Field Archery World Championships. This is important, because it goes to show that Henry has very good aim. So good, in fact, that Henry can target some of the tiniest parts of the body, down to the level of cells and viruses. To be fair, though, it's not his archery skills letting him do that, it's his innovation in biotechnology – a system called KODE. Henry is a proud Kiwi – the original name of his company was Kiwi Ingenuity, but the American customers thought it had something to do with fruit, so with a friend he doodled some names on the back of a

Now you are wondering what's the point of all this? Why stick things to things so much? The point is to deliver the function they want to deliver; the function might be a tiny molecule that acts as an antiseptic – if you've applied the KODE to air-conditioning filters, you've now made them also kill bugs.

100

KODE Biotech have applied for over 100 patents and have an impressive line-up of customers, including the University of Oxford. They foresee this business being worth tens or hundreds of millions of dollars.

doily, and landed on Kiwi Ingenuity; Ortho Designer; Erythrocytes (KODE) – which in his scientist mind translates as 'custom-designed red blood cells'.

To help understand the cleverness that is his KODE system, imagine for a minute that Henry shot one of his arrows at the front door of your house. Imagine the tip of the arrow sticking into the wood of the door, the shaft of the arrow sticking out from it, with the feathers at the end of the arrow (the proper name for them is 'fletchings' by the way). Imagine on the fletchings you've attached a new number for the house – number 19. Now from the street, the postie looks up and sees both numbers – your original number, and this new one stuck to the door. Fooled, they deliver both sets of mail to you and move on, whistling as they go.

Take all of that, shrink it down until it's so tiny only a powerful electron microscope can see it, and that's essentially KODE. It's a system for delivering a payload (the number) to almost any surface at a very, very small scale. The system can work on virtually any surface – the arrowhead, which is actually a thing called a lipid, will stick into glass, metal, plastic, rubber, fabric, cell membranes, viruses – and the beauty is that Henry's technology doesn't really have to aim. The scientists can mix the substance with water, apply it to the surface, and the tiny molecules of KODE will stick within seconds. They can even put the KODE into a printer and print them onto paper. The original substance – the cell or virus or other organism you have now stuck the KODE onto – is unchanged except it now also has this new 'Koded' function. It hasn't been done before like this, and the system is so good at sticking to stuff that it surprises even the inventors. 'We never thought from the beginning it would be able to do what it can do; every time we tried to fail, it succeeded.'

Now you are wondering what's the point of all this? Why stick things to things so much? The point is to deliver the function they want to deliver (the thing stuck on the other end of the arrow). For example, the function might be a tiny molecule that acts as an antiseptic – if you've applied the KODE to air-conditioning filters, you've now made them also kill bugs. The payload might be a specific chemical signature – you've now micro-encoded the substance at a level no one will ever be able to forge. It may be a substance that will light up (phosphoresce) when it detects a certain virus or bug – you've now created a sophisticated bug detector.

Or take a medical example – you might not know this, but there are some aspects of human blood types that change according to race. Asian-heritage people's blood group systems have slightly different properties than those of people of European origin. The KODE can be used to make European blood appear to be Asian blood, for things like testing blood transfusions, stopping rejection of new blood cells, and other purposes.

Some people describe the KODE system as like painting the substance to make it appear like something it isn't. The uses of the technology are extremely varied, and Henry and his team are still coming up with unique applications in industry, medicine, drug delivery, and even cosmetics. It can be used for attacking cancer cells, or protecting newborn babies. KODE Biotech have applied for over 100 patents and have an impressive line-up of customers, including the University of Oxford. They foresee this business being worth tens or even hundreds of millions of dollars; and as a private enterprise with Auckland University of Technology as a shareholder, they stand to be globally successful. But according to Henry, 'Inventing is the easy part; commercialisation is the hard part.' They have to turn this clever idea into a sustainable business, and a research-intensive company like theirs sucks up money in the early days. After more than 16 years of work, they are only now seeing profits ahead of them. In 2013 they signed a partnership agreement with a US company to license their technology and act as an agent. They're also selling online, and working with other possible agents and business partners around the world.

The Variable Room Acoustics System

SOUND RESEARCH AND SCIENCE LEADS TO COMMERCIAL SUCCESS

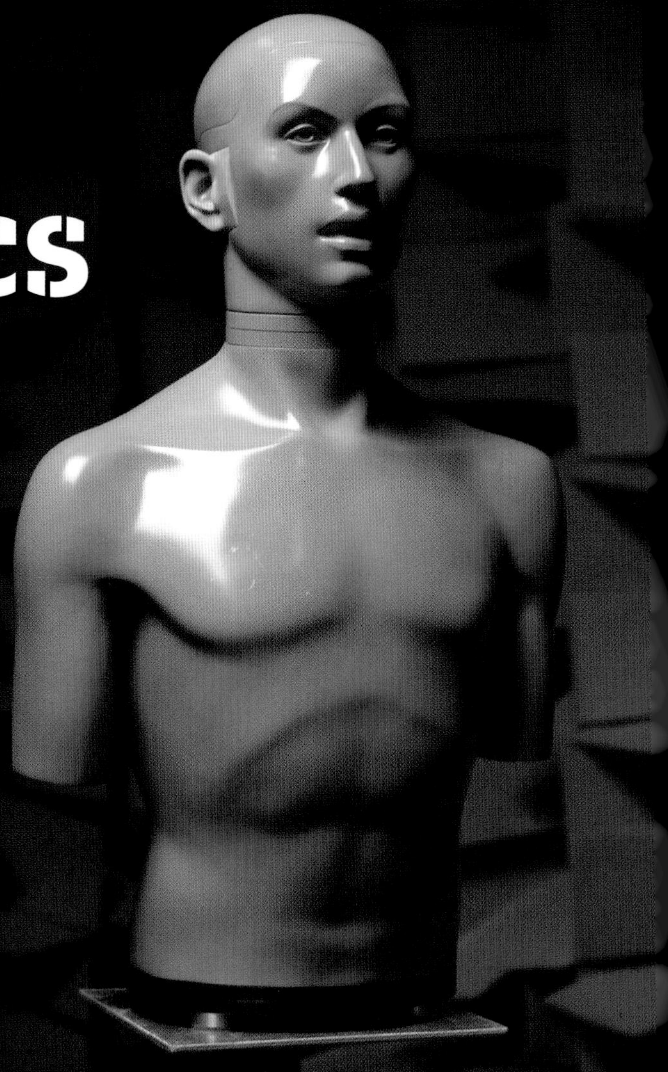

In 1990 Auckland got itself a whiz-bang new concert venue, the Aotea Centre. It had everything: flash foyers, plush seats, a line-up of international artists and orchestras – and really bad sound. The acoustics were so bad that one critic described it as like listening to a mono LP. The sound was flat and with none of the sustained reverberation that makes for great acoustics in a concert venue. It was a dog – with a horrible bark. Unfortunately for the audiophiles of Auckland only a major refit was going to help, and the audiophiles of Auckland had to wait 20 years until one was warranted. One option to remedy the noise was to literally raise the roof five metres – clearly not a viable approach. The alternative was to install the brainchild of Industrial Research Ltd scientist Dr Mark Poletti.

Poletti specialises in acoustics and the science of sound. Among his many innovations are a

20

Only a major refit was going to help, and the audiophiles of Auckland had to wait 20 years until one was warranted. One option to remedy the noise was to literally raise the roof five metres. The alternative was to install the brainchild of IRL scientist Dr Mark Poletti.

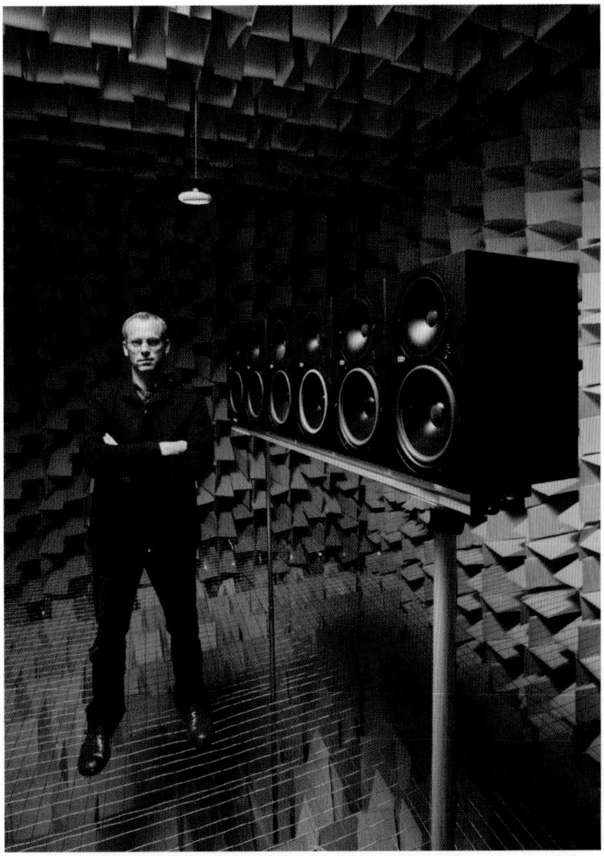

Right: Dr Mark Poletti testing his acoustics system.

solid-state guitar amplifier ('If you want something done properly, do it yourself!') and a number of patents around the design of algorithms for sound processing. It was one of these, the variable room acoustics system, or VRAS, which became a commercial hit. He designed the system using multiple microphones to enhance and control the reverberation of the room. The mikes make it seem like the walls are reverberating more than they are, effectively mimicking the effect of a much larger space and acting like an electronic extension to the walls. So the VRAS system is not really an amplifier as such, it's more like a way to electronically make the room *feel* bigger and give it more reverberation.

The VRAS algorithm and system were licensed to a US partner and eventually bought by Meyer Sound, who incorporated them into a key product line, the Constellation. In this way Poletti's invention has achieved high-profile success including installation at the University of California, Berkeley – as well as the Aotea Centre.

Poletti still works in the area of sound and acoustics. He's currently working on 'holographic audio', which is almost as sci-fi as it sounds. Basically

He designed the system using multiple microphones to enhance and control the reverberation of the room. The mikes make it seem like the walls are reverberating more than they are, effectively mimicking the effect of a much larger space.

he's inventing ways to fool the ears into hearing sounds in places they aren't coming from. He's also helped New Zealand company Phitek with their sound systems for planes. He continues to be an innovative world leader in his field.

Now when you attend a concert in Auckland, chances are the sound waves hitting your inner ear have been subtly enhanced by Poletti's algorithm – adding depth to the sound and, in the words of that opera critic, 'volume, bloom and the stereo-like envelopment you get in a great lyric theatre'. Now, if a dog were to bark in the Aotea Centre it would sound great.

Software

CLEVER TECHNOLOGY BRINGS BIG BUCKS

Computer software innovation can be hard to pin down – indeed, the recent changes to New Zealand patent laws specifically exclude software patents. It's also hard to tell what's an innovation, versus an evolution of a previous idea. That's not to say the New Zealand software industry is not doing well – indeed, it's humming. The independent commentator Technology Investment Network (TIN100) report on the industry consistently shows that information and communication technology (ICT) contributes hugely to the New Zealand economy, and to our exports. New Zealand digital companies are world leaders in gaming, application development and healthcare.

This is one industry where the tyranny of distance is almost irrelevant. The barriers to success in software are more typically around capital – software takes a lot of effort to create – and accessing the right sort of international expertise or market advice. Initiatives like the Kiwi Landing Pad in San Francisco attempt to get software tech entrepreneurs closer to the heart of the action, and it seems the primary benefit is to expose them to the stimulating environment of Silicon Valley, with its fast pace and its merit-driven ethos. The Valley doesn't care where you come from, where you live, or who you are – if you've got a good idea and the team to deliver it, you can succeed. Rather than attempt to document all of the successful Kiwi software companies, here are a few choice examples of firms and entrepreneurs who have created extremely successful and influential software inventions.

From zero to hero to Xero

Today Rod Drury is pretty much a household name. His avuncular visage is a regular feature on the telly and every time someone wants to gush over an internet success story they choose Drury. And it's no surprise when you know that he's built, grown and sold a number of internet businesses to great success. He set up a locally focused IT company in the mid-90s, then created an email archiving product (Aftermail) which he successfully sold to a US software company for a reported US$65m. Instead of taking the cash for the classic 3Bs – bach, boat, BMW – retirement many tech millionaires choose, he went from hero to Xero, establishing the online accounting software company in 1996.

Determined to grow a billion-dollar software company, Drury has driven the company with incredible passion and seen great success on the stock market, if still not convincingly in the sales results. At the time of writing Xero was New Zealand's most valuable listed company by market capitalisation. In all of these companies, Drury has innovated, both in product and approach, and shown that New Zealand internet entrepreneurs can foot it with the best of them.

Ghost

Imagine a large US company knocking on your door and offering you a fortune to buy your idea. That's pretty much what happened to Auckland computer programmer Murray Haszard. His product, Ghost, was an example of having the right idea at the right time – and having the right amount of luck to go with it. Haszard had created a programme to clone hard disks for IBM's operating system OS2. Yes, the one you've never heard of. OS2 didn't really sell and neither did Ghost, but luckily for Haszard in 1995 Microsoft launched a new version of their Windows operating system, and was delighted to discover that his product would work (with a few minor tweaks) on that platform too. Not only that, but all the pre-existing products from the Windows and DOS world wouldn't! He was in a league of his own. He set up a website to sell Ghost – one of the first to sell in this way – and Haszard says it was like 'magic'. The sales took off exponentially, doubling each month until they were getting revenues of about US$2m per month. Early in 1998 they had a real offer from Symantec, and in 1999 the sale was finalised. Haszard gave Symantec the rights to his product and in return they gave him a cool US$27.5m.

Wildfire

Victoria Ransom is from Scott's Ferry, near Bulls, and she is pretty much the biggest thing to come out of Bulls since they stopped the A&P Shows. In 2007 Ransom created a product and set up a company, Wildfire, with her partner Alain Chuard. Wildfire is an internet social media marketing firm which allows companies to promote themselves on sites like Twitter and Facebook. True to its name, the company took off like wildfire and in 2012 captured the attention of internet giant Google, which bought her company for a reported $250m, surely making her one of New Zealand's most successful business people.

Weta Digital

In 1993 Weta Digital was created by Peter Jackson, Richard Taylor and Jamie Selkirk. Taylor (now Sir Richard) was a props maker who'd worked with Jackson on *Meet the Feebles*. Selkirk was a former film editor who had been with Jackson since *Bad Taste*, and Jackson (now Sir Peter) was...well we know who Jackson was and is. The company was formed to take on the challenge of fulfilling the ambitious digital effects for *Heavenly Creatures*, and since then it has been at the very forefront of digital effects technology. With a philosophy that today's innovations are tomorrow's limitations, Weta has won five Academy Awards and made some of the biggest films of all time.

And they've made plenty of inventions in this esoteric digital space. The proprietary software they've created to make both smaller detail and larger scale seem real constantly advances the art.

James Cameron waited 10 years for digital arts technology to be up to the task of creating *Avatar*. When someone finally caught up to Cameron's vision, it was Weta. Their Virtual Studio was a world first.

Actors who are going to be mapped onto digitally created bodies – such as *Avatar*'s Navi – are filmed moving and speaking in a motion-capture studio called a volume. It's just a large empty room, so when the actors are acting, there is nothing for the director to look at but bodies in a blank room. Cameron challenged Weta Digital's Glenn Derry to come up with a way for him to see the actors set into the background on the planet Pandora as they acted. He needed to be able to move a virtual camera around them and see them from different angles in this virtual 3D world.

The result was the swing camera and Virtual Studio. Cameron can swing the camera around the actors. The camera senses where it is relative to the actors. Computers then superimpose the actors on the digital environment and the resulting low-resolution pictures are fed to Cameron live in the studio. He can then watch the actors on Pandora, seeing how their action works in context.

Hotdog

It's what every computer geek dreams of – coming up with a cool idea and making millions from it. The history of computing in the last 20 years seems littered with examples of geniuses beavering away in their bedrooms for weeks on end and then becoming instant millionaires. Kiwi Steve Outtrim was one such geek for a while. Outtrim created an editor for web pages, Hotdog, and launched it in 1995. In the very first week of operation he made over $3000 in sales, in the first month over $36,000, in the first year over $3 million! Sausage Software, the company he founded in Australia, became a darling of the new dotcom boom. He took the company public and became the youngest CEO in Australia to do so. Sausage Software is now part of another company, and Outtrim has moved on to other endeavours, but for a while at least, he was the top hotdog.

Others

There are myriad other successful software developers: – Small Worlds, who have created a hugely popular online gaming platform; Grinding Gears, whose game *Path of Exile* has made millions online, despite being free to play; Orion Healthcare, whose software products manage hospitals all over the world; Vista Entertainment, world leaders in cinema management software; Massive, the software platform that was created for the crowd scenes of *Lord of the Rings* and now is used in numerous ads and feature films; the other Massive, the software sold to Microsoft by Kiwi Claudia Batten, which can insert advertising into games; Peace Software, who developed billing systems for utilities; Beetil, a tool for managing best practices in IT companies, which was bought out by Silicon Valley giant Cisco; Vend, creating a world leading point-of-sale system; Jade, who created a truly innovative database platform; Greenbutton, who worked out a way to take all the old servers from the *Lord of the Rings* and create a remote super-computing platform for others to use; Pegasus Mail, one of the first email servers; and many more...

Designer Endophytes

GREAT SCIENCE PREVENTS AGRICULTURAL DISASTER

1991

They pioneered a way to create brand-new endophytes that would protect our grass from pests but not cause staggers, and then in 1991 they patented a way to inoculate the grasses with these new endophytes.

Right: Micrograph of stained fungus within plant tissue.

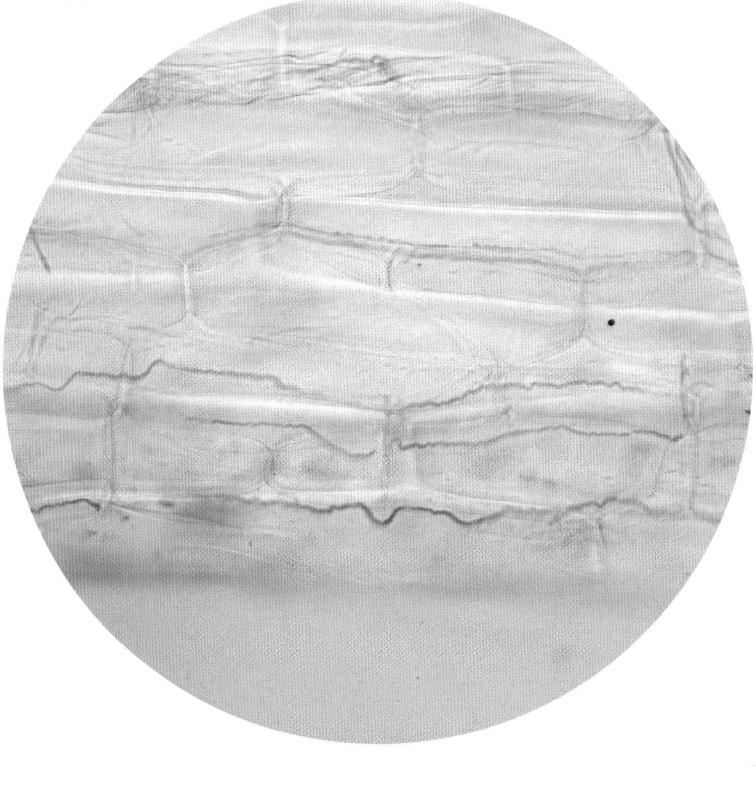

New Zealand's most important commodity is grass. We've been using animals to mine our grassland riches and sustain our economy for most of our history, and we're still largely dependent on these billions of green strands. If anyone is going to invent a microscopic bug to increase grass production, it's New Zealand.

To tell this story I need to introduce you to a cast of characters. First there's our hero: ryegrass – our most common pasture. The bad guys are the Argentine stem weevil (evil), root aphid (rude), black beetle (racist) and pasture mealy bug (disgusting). The hero's sidekick in this adventure is the endophyte.

In the early 1970s New Zealand's AgResearch scientists began to investigate a mysterious and awful condition called staggers, where an animal trembles, loses condition, and can stagger or fall over. You might argue that those could just be particularly clumsy sheep and cows, but in 1981, after some solid research, wonderful accidents and brilliant deduction, scientist Lester Fletcher pinned staggers once and for all on endophytes in the grass – which is probably why staggers is now commonly known as grass staggers. Endophytes – fungal parasites that live their entire life cycle inside the grass itself – had long been known but this discovery was a first.

2000

Field testing proved very successful and in 2000 AgResearch released the world's first novel endophyte, AR1. This little invention was the result of 20 years of research by a team of mycologists, chemists, agronomists, entomologists and horticulturalists.

Left: AgResearch scientist Garry Latch and senior technician Anouck de Bonth.

New Zealand's most important commodity is grass. We've been using animals to mine our grassland riches and sustain our economy for most of our history. If anyone is going to invent a microscopic bug to increase grass production, it's New Zealand.

You'd think this would cause a massive endophyte backlash, but around the same time the good folk at AgResearch also discovered that endophytes gave grasses resistance to the cast of nasty pests mentioned above – pests that have the power to devastate our grassland.

AgResearch and its commercialisation arm Grasslanz (the 'nz' at the end also stands for New Zealand!) went looking for a smart way out of this conundrum, and in doing so became the world leaders in endophyte technology. They pioneered a way to create brand-new endophytes that would protect our grass from pests but not cause staggers, and then in 1991 they patented a way to inoculate the grasses with these new endophytes – a technique its inventor Garry Latch called 'slash and stuff'.

Field testing proved very successful and in 2000 AgResearch released the world's first novel endophyte, AR1. This little invention was the result of 20 years of research by a team of mycologists, chemists, agronomists, entomologists and horticulturalists. The subsequent improvements to New Zealand's pasture and animal health are probably worth hundreds of millions of dollars a year to the country.

And stock growth isn't the only potential for endophytes. What about bird strike at airports – a $1.2 billion problem? AgResearch scientist Chris Pennell was flying in to London when his plane was hit by bird strike. He began to think about the problem. Using two new endophytes, AR601 and AR95, Pennell's team made a special seed to control birds at airports around the world. The grass is resistant to insects and repugnant to birds – so the birds stay away. PGG Wrightson are selling this remarkable invention internationally under the name Avanex.

To date AgResearch and Grasslanz have developed, patented and commercialised many new endophytes and are selling them all over the world. Each of these little sidekicks offers different ways of helping the hero grass, and all of them are little New Zealanders.

Wave Power

HARNESSING THE ENERGY OF THE SEA

With all the talk in this book about saving energy with more efficient engines and motors, and different ways of using energy like wireless power and charging, there should be a few words about New Zealand's innovative efforts to produce energy.

The world's scientists are sprinting full tilt in a high-speed technology race to harness the energy of the oceans, the movement of which contains enough power to satisfy the needs of humans many times over. But building generators that can handle the destructive and corrosive power of the sea and give reliable, efficient energy no matter what the wave conditions has proven too hard so far. There are a few tide and wave energy projects around the world but they are still relatively small scale, and the wind and solar industries are about 15 years ahead.

New Zealand also had early wave power projects – in 1903, Dunedin man Robert Miller demonstrated his wave power device to a crowd of prominent citizens. He had put a pontoon in the middle of a small cove, tied to the coast with wire rope. As the waves rose and fell, they'd pull the rope and back on shore a 'power machine' would drive a flywheel to about 1 horsepower of output. He calculated that the power output was proportional to the weight of the pontoon and that, if he had one big enough, he could get up to 500hp from the device. A group of locals liked the idea and backed it, patenting the approach, but ultimately the idea faded as the practical implications seemed too hard to overcome.

In 2004 Alistair Campbell and his team at Callaghan Innovation started to consider the process in a

2004

In 2004 Alistair Campbell and his team at Callaghan Innovation started to consider the process in a different way to other researchers around the world.

Left: The wave power device, quietly harvesting away.

1

If I was a betting man I'd be putting a dollar each way on this particular Seabiscuit but the fact is this technology doesn't have to win, it just has to get to the finish line: commercially viable power from the sea.

Below: The system rights itself when waves displace it.

Most wave power generators capture either the up-and-down motion of the waves (heave) or the back-and-forward motion (surge). What's clever and different about this Kiwi approach is that their massive float spins between its uprights, capturing not only the heave but also the surge.

different way to other researchers around the world. As engineers rather than mathematicians they looked to develop an engineering solution that would be robust enough not to get smashed and sink – the fate of most early attempts. Says Campbell, 'We started dreaming stuff up in the shower, watching the water flowing.' While their overseas rivals were testing small models in wave pools, Campbell and his team loaded their models onto trailers and tested them in their wave pool – Lyttelton Harbour. And they were able to prove that not only did their eventual solution deliver power, it survived the ocean.

Most wave power generators capture either the up-and-down motion of the waves (heave) or the back-and-forward motion (surge). What's clever and different about this Kiwi approach is that their massive float spins between its uprights, capturing not only the heave but also the surge.

Originally called WET-NZ, the invention is now licensed to an American company called Northwest Energy Innovations (NWEI) who, along with New Zealand company EHL Group, carried out successful testing in Oregon in 2013 and are testing it with the US Navy in Hawaii in 2014.

If I was a betting man I'd be putting a dollar each way on this particular Seabiscuit. But the fact is this technology doesn't have to win, it just has to get to the finish line: commercially viable power from the sea. If it can do that, it'll pay off big for NWEI, EHL, Callaghan Innovation and New Zealand.

Stem Cell Pathways

A MOTORWAY OF NEW CELLS IN OUR BRAINS

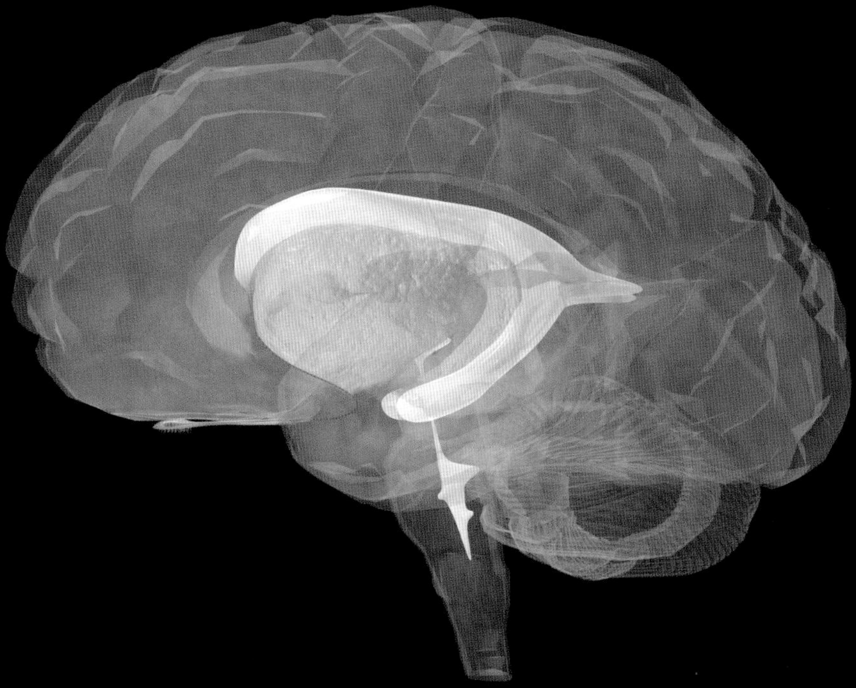

The 2007 recipient of the Rutherford Medal was Professor Richard Faull, a man who keeps a 'brain bank' of 400 brains at work, and carries around a real human brain, hardened with resin, everywhere he goes. That might seem weird or macabre but Faull's 30 years of world-leading research led to his discovery of extraordinarily good news about the human brain.

Starting life in Tikorangi, Taranaki, Faull saw his first human brain at medical school – 'It was love at first sight,' he says. After he graduated from medical school, Faull went into neurosurgery, presumably to stay close to his love. But when he discovered that we really know little about the brain, Faull gave up surgery and got into research.

The accepted scientific opinion was that the human brain is degenerating every day and that it has no mechanism for regeneration. It's a pretty sorry prognosis. Over 15 years, Faull built up a bank of brains donated to him by families of people with brain diseases such as Huntington's. Says Faull, 'We found all different things going on in the brain that

30

Faull's 30 years of world-leading research led to his discovery of extraordinarily good news about the human brain.

Left: Professor Faull, a man with a lot of brains.

Stem cells are brain cells that are able to divide into any sort of brain cell so are like a tool kit for repairing the brain.

were not in the textbooks.' One unexpected finding was that brains with Huntington's disease seemed to be making new cells, trying to repair themselves. This was extremely controversial as it was against current thinking.

Of course the ultimate aim of studying brain disease is to find a cure: says Faull, 'The dream is to replace the diseased cells.' So, after years of eureka moments in understanding the brain, Faull and his team deliberately set out to smash the belief that the human brain doesn't regenerate.

It had been known since the 1960s that the brains of rats and other animals have a pathway which transports brand new stem cells to their brain's olfactory (smell) centre. Stem cells are brain cells that are able to divide into any sort of brain cell, so are like a tool kit for repairing the brain. After many years of failed attempts to find such a pathway in our own brain, scientists believed it did not exist. It seemed like rats had more chance of fighting off brain disease that we do – unfair.

Faull and his former PhD student Dr Maurice Curtis together with collaborators in Sweden took human brains and painstakingly cut them into thousands of extremely thin cross-sections. Their breakthrough was that they cut the cross-sections on different axes than others before them had. After three years' work they found the motorway, and they carefully mapped it across the brain. They wrote up their discovery and sent it to the leading scientific journal, *Nature*, where the referees rejected it because they didn't believe that a team of Kiwis could have found what nobody thought existed. Undaunted they sent their findings to *Nature*'s rival *Science*, who not only published it in March 2007 but featured it on the front cover. 'It was even better than being the centrefold of *Playboy*!' says Faull 'When Galileo suggested the Earth was not the centre of the universe, the Church put him under house arrest. We're lucky we're still free!'

Faull's research continues today. He is part of a massive, collaborative worldwide research effort on the human brain. 'Everything which human endeavour strives for, we achieve through brain activity. I personally believe the greatest challenge to scientists – and one of the last challenges – will be to unravel what makes it work.'

Until then, Faull recommends keeping your brain young with exercise and creative thinking.

Medical Honey

HONESTLY, SMEARING HONEY ON YOURSELF IS 'SCIENCE'

1974

Stratford foresaw that honey had uses beyond toast: he imagined a medical use for it, and in 1974, with co-founder and businessman Alan Bougen, he set up Comvita.

Right: Claude Stratford and his smokepot.

Tauranga company Comvita was founded, they're proud to admit, by a hippy with a mission. Claude Stratford (b. 1910, d. 2013) was a 'kooky' visionary in that he saw a new use for honey. Stratford foresaw that honey had uses beyond toast: he imagined a medical use for it, and in 1974, with co-founder and businessman Alan Bougen, he set up Comvita (the name implies life). Their idea was to create a business that could help people in the community – they were very early adopters of the social capitalist mind-set that is becoming more popular these days. They thought that the health paradigm at that time was backwards – it focused more on illness than wellness – and they set about reversing that. With honey.

From their early days, the idea of a scientific approach was part of the ethos – the full name they called their business was Comvita Laboratories Limited. Early products included herbal creams, and although the formulations may have had some unusual ingredients – 'Some of Freda's roses, when in season, bless her' – hippy Claude followed a fairly scientific method to check for effectiveness. Business was tough and developed slowly. Selling food ingredients and edible honey kept them afloat. Then in the late 1980s the world started to catch up

Above: A native New Zealand bee inside a mānuka flower.

They were very early adopters of the social capitalist mindset that is becoming more popular these days. They thought that the health paradigm at that time was back-wards – it focused more on illness than wellness, and they set about reversing that. With honey.

with them. A worldwide trend towards healthy food increased interest in their ideas. A new area called apitherapy popped up and Comvita were perfectly placed to jump on this new approach.

They increased investment into researching the applications of honey, and started to make world-leading contributions in the area. They put some serious science behind what the hippies had been saying for a while – that the honey from bees that pollinated the scrubby mānuka plant was somehow magical. Mānuka was cosidered a weed tree, but Comvita's scientists helped show that mānuka flowers contained active chemicals that were unique and had scientifically valid health benefits. The bees collect it for free, and so it goes freely into their (luckily delicious) products. They call this the Unique Mānuka Factor (UMF) and it has meant their products are sought after for their health benefits. It allows them to capture a lot more value – read: make more money – from the same source ingredients. Today Comvita are a $100m-plus world leader in the area, with patents and trademarks such as Medihoney.

In an interesting loop back to their past, their research has helped extend the use of honey products from being wellness products: they are now also used to treat wounds, burns and various illnesses. Again, the science backs up what the hippies were saying all along – honey is more than just a food, it's a lifestyle.

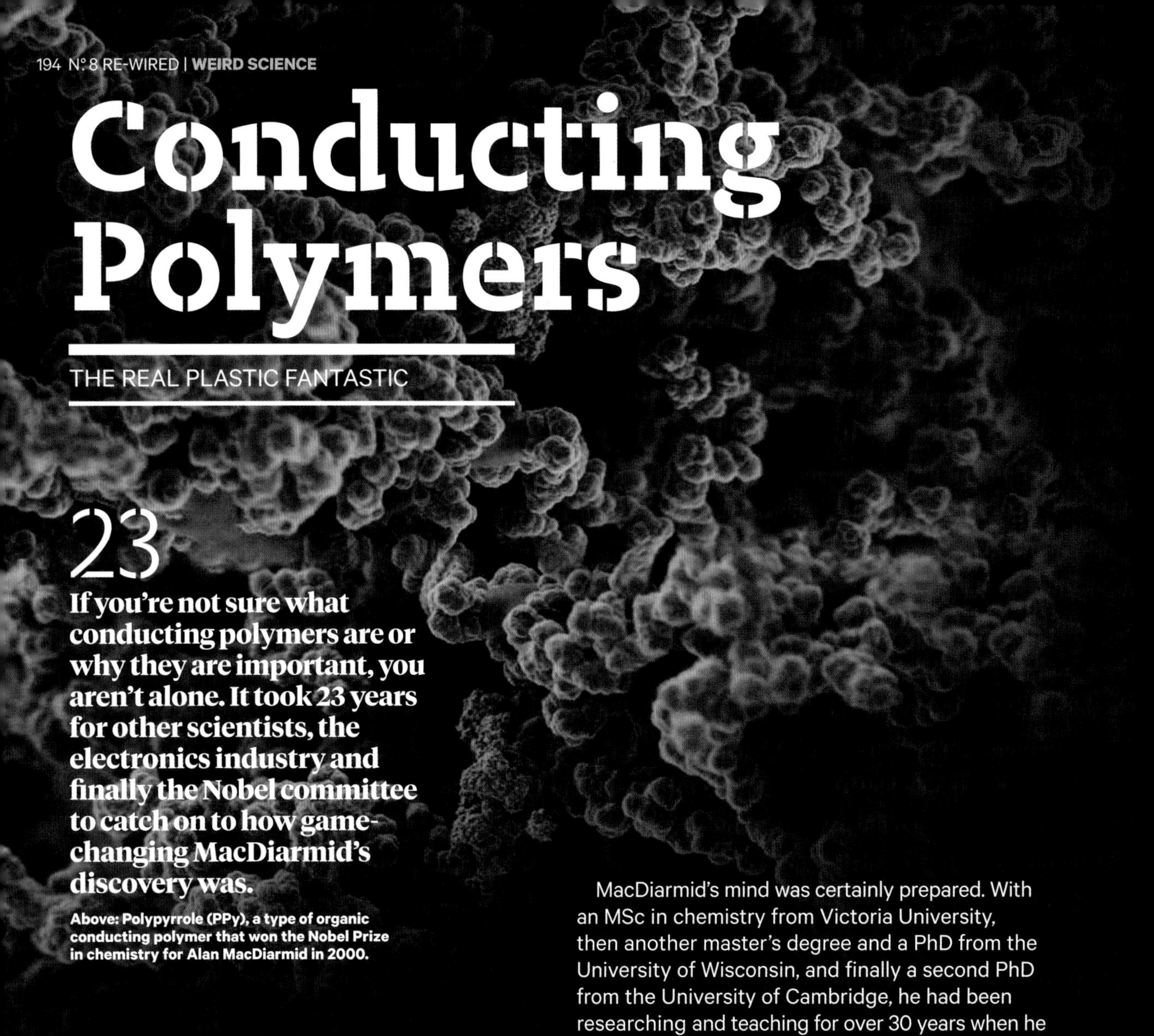

Conducting Polymers

THE REAL PLASTIC FANTASTIC

23

If you're not sure what conducting polymers are or why they are important, you aren't alone. It took 23 years for other scientists, the electronics industry and finally the Nobel committee to catch on to how game-changing MacDiarmid's discovery was.

Above: Polypyrrole (PPy), a type of organic conducting polymer that won the Nobel Prize in chemistry for Alan MacDiarmid in 2000.

The Nobel Prize is the most prestigious award that a human can receive. Out of about 850-odd Nobel laureates in just over a century, there have been only three New Zealand recipients. One of them described the discovery that led to his prize like this: 'If you go to Las Vegas or Atlantic City and put a quarter in the slot machine, a vague thought goes through your mind that maybe you might hit the jackpot. Similarly, if you go into a lab and play around with new things, the thought vaguely goes through your mind that maybe you might hit a scientific jackpot. But it's not something you really consider seriously.'

Of course Alan MacDiarmid (b. 1927, d. 2007) was being modest. His scientific discovery, like penicillin or Teflon, may have had an element of luck, but as Pasteur said, 'Chance favours the prepared mind.'

MacDiarmid's mind was certainly prepared. With an MSc in chemistry from Victoria University, then another master's degree and a PhD from the University of Wisconsin, and finally a second PhD from the University of Cambridge, he had been researching and teaching for over 30 years when he 'got lucky'.

The breakthrough that earned him the Nobel Prize? Conducting polymers.

If you're not sure what conducting polymers are or why they are important, you aren't alone. It took 23 years for other scientists, the electronics industry and finally the Nobel committee to catch on to how game-changing MacDiarmid's discovery was. Put simply, a conducting polymer is a plastic that can conduct electricity. Before MacDiarmid and his collaborators published their findings in 1977, it was thought the two terms were mutually exclusive.

To help illustrate why the discovery has had such an impact on the world, look at it this way: On the one hand we have plastics, which are easy to synthesise, cheap to process, light in weight, and readily mouldable into any shape. We all know how many

The discovery of a way to make plastics conduct electricity. threatens to be as revolutionary to electronics as the transistor or semiconductor were.

uses plastics have in everyday life. On the other hand we have something equally important – metals that conduct electricity. Conductors are elemental in the manufacture of all of the electronic devices that are increasingly changing our world. But conductors are all metals. They are neither cheap, nor easily synthesised, easily processed or light in weight.

The discovery of a way to make plastics conduct electricity and bring the advantages of plastics to electronics threatens to be as revolutionary to electronics as the transistor or semiconductor were. It could lead to a complete change to the way electronic devices are made – giving us the very real possibility of things like wallpaper that lights a room, flexible computer screens that you roll up and pop in your bag, a laptop the size of a watch, or solar paint that covers your house and at the same time is one big solar panel. Whatever revolutionises electronics has enormous power to change the way we live, and that is what conducting polymers promise.

MacDiarmid's great breakthrough began in 1975 over a cup of green tea in the Tokyo Institute of Technology with Japanese scientist Dr Hideki Shirakawa. Shirakawa had a foreign student working for him who had misunderstood the Japanese measurement in a preparation he was making, and added 1000 times too much catalyst to the mixture. Sitting in the café, MacDiarmid was showing Shirakawa a golden-coloured inorganic polymer he had discovered could conduct electricity. Shirakawa said, 'I have something like that,' and showed MacDiarmid the results of his assistant's mistake – some simple polyacetylene that had turned all silvery. Shirakawa's polymer also conducted electricity, albeit weakly. The two scientists realised there was potential for research in what they had each found.

MacDiarmid applied for and received funding to invite Shirakawa to the University of Pennsylvania for a year to investigate this idea that plastic polymers might be made to conduct electricity. They joined forces with a University of Pennsylvania physicist called Alan Heeger and set about it.

Polymers are long chains of organic compounds. Many of them like polythene and PET plastic are very familiar to us. They don't conduct electricity because electricity is conducted by free electrons within a substance and polymers have no free electrons – all their electrons are bound up in the bonds that join their molecules together. MacDiarmid and his collaborators found that if they 'doped' certain polymers with an impurity of the right sort, it could free up electrons all along the chains of the polymer, thus allowing the polymer to become conductive while still retaining its plastic properties. Said MacDiarmid, 'At first people didn't necessarily believe what we were saying, but that slowly changed.'

Their findings opened a floodgate worldwide. MacDiarmid, Shirakawa and Heeger found themselves the fathers of a whole new field of scientific research into conducting polymers.

Some practical applications are already in the marketplace. Organic LED TVs are the must-have gadget at time of writing. These 'OLED' TVs take advantage of the discovery in 1990 that some conducting polymers emit light when a voltage is applied to them.

And television is just one way we can expect to see MacDiarmid's invention entering our homes and our pockets. The opposite of an LED is a solar cell, which collects light and emits electricity. Companies, universities and institutions around the world are frantically researching organic solar cells in the hopes of developing something as exciting as a light, flexible and cheap alternative to current solar technology. There are also applications in batteries, in circuit boards, and as a replacement for solder. Wherever expensive copper and silicon are used today, polymer could step in, making things lighter, smaller and cheaper.

Locally, the University of Auckland's MacDiarmid Institute has been set up to forward research into nanotechnology and advanced materials like conducting polymers. The idea is to create a network of scientists across the country and to apply their ground-breaking research by partnering with New Zealand businesses to make new products for export.

It's genuine No. 8 Re-wired, and this time the wires are organic. There's a very real chance that polymers could one day be the new silicon. Imagine if the new Silicon Valley was Polymer Valley – and imagine if it was in New Zealand!

10.

Close but No Cigar

DAYLIGHT SAVINGS, DESIGN WINNERS AND THE AIRTIGHT TIN LID

Without wishing to seem trite, let us remark that failure is an integral part of inventing. Most inventors, and even 'serious' scientists, will say that the road to success is littered with unsuccessful prototypes. Celebrating failure, lifting yourself up from one mess and looking forward to the next attempt appears to be a core trait of successful inventors.

Having said that, some of the stories of inventors are tragic. Despite their hard yards, they don't make it into the select group of inventors who have achieved financial success. No doubt these tragedies are legion. Aviation pioneer Richard Pearse (see page 78) is perhaps the most famous of our invention tragedies. Quite apart from the question of whether he flew first, the fact is that he let his patent on the aileron lapse, just before the international aircraft manufacturing industry adopted it for good – a monumental inventing mistake.

Probably more than half of the inventions in this book have failed, at least to date, to receive the financial recognition that the raw idea promised. The stories we have chosen here are the stories of the best inventions with the poorest returns to their masterminds. They illustrate a key point – a good idea is a start, a good product is better, but to be truly successful, it takes a mindset that perseveres, embraces failure, learns from others, and admits when they need help. That's No. 8, Re-wired.

Daylight Saving Time

DON'T FORGET TO PUT YOUR CLOCKS FORWARD...OR IS IT BACK?

Right: George Vernon Hudson, second from left, with members of the Auckland Islands Sub-Antarctic Expedition team (1907).

To have the audacity to play with the very idea of time itself is surely an exercise of extreme hubris. Yet that's what George Vernon Hudson (b. 1867, d. 1946) proposed to do in 1895 when he suggested to the Wellington Philosophical Society that they consider moving the time around twice a year to accommodate the longer hours of summer.

What a preposterous notion! The idea was initially ridiculed by members of the society. One Mr Maskell said that 'it was out of the question to think of altering a system that had been in use for thousands of years'. A Mr Travers (probably in a sarcastic tone) asked if the idea was to have clocks with two sets of hands; and one Mr Harding said that Hudson's idea was wholly unscientific and impracticable. Mr Hudson replied that 'he was sorry to see the paper treated rather with ridicule. He intended it to be practical. It was approved of by those much in the open air.' And he sloped off to play with his insect collection.

For Hudson was, among other things, an entomologist. It had been this hobby, fitted in around his shift-work job, which had prompted his idea of changing the clocks; imagine if we could move time to enjoy the summer evenings. He wrote in his submission that a 'long period of daylight leisure would be made available in the evening for cricket, gardening, cycling, or any other outdoor pursuit desired'. Clearly Hudson was a visionary – or someone who was hoping to get the entire country to change its clocks so that he had more time to collect butterflies after work.

78

While New Zealand was not the first to adopt daylight saving, Hudson is universally recognised as its inventor. Today, around 78 countries use daylight saving time.

Left: In 1984, the dairy-farming residents of Ararua, Northland, rebelled against daylight saving time.

Clearly Hudson was a visionary – or someone who was hoping to get the entire country to change its clocks so that he had more time to collect butterflies after work.

Hudson was born in England but moved to New Zealand as a teen. Basing himself in Wellington, he wrote a number of books on insect fauna, and travelled on scientific expeditions to explore the wildlife of the subantarctic. He amassed the largest collection of insects in New Zealand (currently held at Te Papa), and as if that wasn't enough, in 1918 he also discovered a new star, Nova Aquilae. Does this sound like a man who would let a good idea drop? No, it does not…

Far from being dead, Hudson's ideas on 'Seasonal Time-adjustment in Countries South of Lat. 30°' started to take seed in the popular imagination, particularly in Christchurch among the, as he put it, 'numerous classes, who are obliged to work indoors all day, and who, under existing arrangements, get a minimum of fresh air and sunshine'.

Thousands of copies of Hudson's original paper were printed and circulated. Cantabrians started to take the whole idea of moving time around seriously. Luckily for all of us summer twilight cricketers, Hudson decided to give the idea another crack, and in 1898 he appeared before the stuffed suits of the Wellington Philosophical Society once again, with an updated, expanded and now well-supported paper. In his second appearance, Hudson acknowledged the critics and their arguments, but presented clear counter-arguments and persuasive reasoning for his case – it was a masterful piece of convincing.

While Hudson had suggested a two-hour shift, ultimately politician Thomas Kay Sidey was successful in ensuring we adopted a one-hour 'daylight saving time' (DST) adjustment in the summer of 1927. It took a long time to convince people that this new innovation would not unduly mess with people's body clocks, or fade the curtains more!

But we were too late to be first – starting at the end of April 1916, Germany and its allies Austria-Hungary began using DST, in part to conserve coal during wartime – it's unclear whether Kaiser Bill nicked the idea from Hudson or thought of it independently. Britain and most of its allies soon followed suit. Russia and a few other countries waited until the next year and the United States adopted it in 1918. While New Zealand was not the first to adopt daylight saving, Hudson is universally recognised as its inventor. Today, around 78 countries use daylight saving time.

Interestingly, daylight saving time continues to be a controversial topic, with some countries (and states within countries – I'm looking at you, Queensland) eschewing it completely, and others re-examining its usefulness. In the US, there is evidence that energy savings are minimal, and in fact that daylight savings and the changes in sleep cycles may mean lower productivity and increased risk of heart attacks. Sounds like scaremongering to me, by those jealous of our own Father Time.

The Power-beat Battery

IT SEEMED LIKE A GOOD IDEA AT THE TIME

For a number of years, Powerbeat stood clearly in the front ranks of New Zealand's innovative companies. Probably their best known invention was the Powerbeat battery, which cleverly separated out the section of the battery that powered a car's starter motor. This meant that even if you left your lights on all day there would still be charge left to start the car and recharge the battery. Brilliant, right? It promised to be a revolutionary breakthrough, but to bring an idea like this to market, to unseat the hundreds of millions of batteries in cars all over the world, it had to be perfect. It costs an astonishing amount of money to take a working prototype to the stage where manufacturing can start. Press releases from the company typically predicted financial success which never seemed to come about. Other inventions that seemed to show promise also did not managed to cross the chasm from R&D prototype to full commercial product, and after years of trying, being listed and then delisted on the stock market, issues with production, issues with sales, the company was wound up in 2011 and an idea that seemed so intuitively right died.

Fly by Wire

NOT QUITE IN THE SWING OF IT

Imagine a swing. Now imagine that it is hung from only one rope so that it can swing in big figure-eights and in all directions. Now imagine instead of sitting on it, you are lying, facing forward, in a little plane. Now imagine the plane is powered by an engine, there's a propeller behind you, and you can steer with a rudder. Imagine all the people coming to New Zealand to ride this powered swing. This is the very train of imagining that Neil Harrap did while lying awake at four o'clock one morning in 1994. So he built one.

The plane can go up to about 170km/h, up to 100m in the air, and down to about 1.5m off the ground. Maximum g-forces are about 3g – about three times gravity – and the pilot has a throttle and steering, so they can fly fast or slow in any direction. It was fairly popular initially in New Zealand, and they had a crack at setting up overseas, but it was not the wild success that, for example, bungy jumping became. The company set up a site in Texas but a big storm forced them to close a few weeks later. The New Zealand operation is struggling too: there was a crash at the Queenstown site in 2001, and both it and the Paekakariki site have now closed. Harrap has some other options for the ride – Turkey is looking good – and some other ideas for new rides too, so no doubt we'll hear him flying overhead again soon.

The Airtight Tin Lid

IT'S ENOUGH TO MAKE YOU CRY

Below: One fail and one success – Eustace's tin lid adorns another Kiwi icon, Edmond's baking powder.

Let's cut right to the chase and state that some thieving foreigners stole a great invention from a Kiwi pioneer and left him without a bean. And all because he didn't get a patent...

Consider Milo. No, hang on, let's make it golden syrup. Or paint. Let's choose paint. In actual fact all of these things have one thing in common – their lid system. The lid on each of these tins is similar, a smaller round plate with a rim that fits snugly into the tin creating a tight fit – airtight, in fact. And this lid system was invented by a Dunedin man, John Eustace.

In 1884 Eustace had a small tinsmithing business, manufacturing, among other things, lids for tins. Back then they were making slip-on lids, which are fine for many purposes, but are sadly lacking when it comes to any application where the contents of the tin need to be protected from the air, such as foodstuffs. Eustace and his brother played around with a few ideas, and struck upon the lid system we know today. Pleased with themselves, they started to manufacture the lids and patented the idea. But here's the rub. They only patented it in New Zealand, and one of the quirks of the patent system is that you need to get a patent in each country you want protection from copycats.

To facilitate the manufacture of the lids, Eustace decided to get a 'die', or cut-out, made in England. The English were obviously very taken with Eustace's ideas, for not long after his dies were made and returned, the pair noticed that some of the paint they were getting from England had their lid design. They were legally unprotected, and could do nothing to stop a flood of companies in England from basically stealing their idea and making money from it.

If only Eustace had realised the importance of protecting his idea with a patent – he could have been a millionaire.

To add insult to injury, one London company contacted the pair and offered them the kingly sum of £15,000 to buy the rights to their design, before realising they didn't have to – they could just start making the lids and there was nothing Eustace could do about it. The company retracted their offer, and started manufacturing millions of the lids.

If only Eustace had realised the importance of protecting his idea with a patent, he could have been a millionaire. It's this kind of experience that certainly fuels the fire of paranoid inventors nowadays. But surprisingly, this bad fortune didn't seem to faze Eustace himself – I suppose he couldn't let it get to him – he went on manufacturing the lids too, growing his business until, by 1927, he was making 100 tons of tin cans per year.

And the remarkable thing is that this Kiwi technology has still not been superseded 130 years later.

The Edlin-Stewart Engine

STEAM STEWART AND HIS DIESEL ENGINE

Below: A detail from the patent application for the engine.

Feb. 4, 1936. G. S. EDLIN ET AL 2,029,990

INTERNAL COMBUSTION ENGINE

Filed June 28, 1930 2 Sheets-Sheet 2

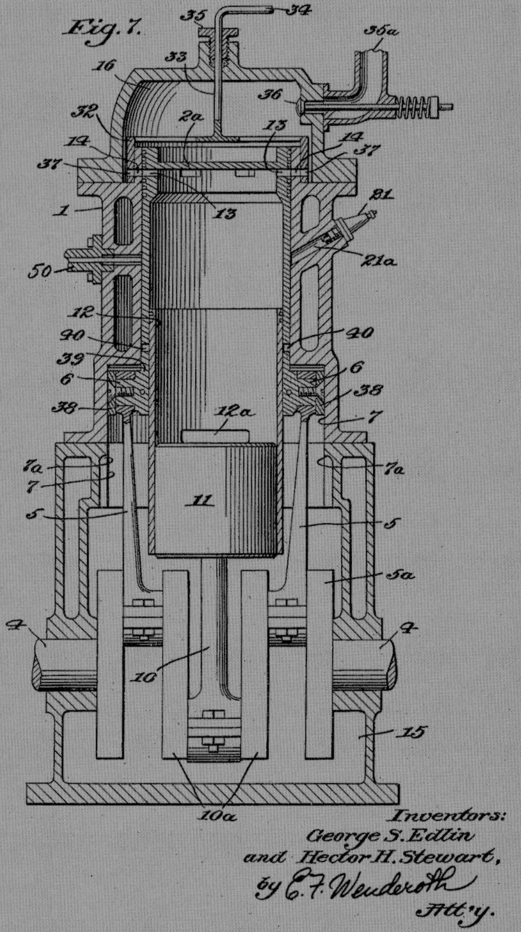

The story of the Edlin-Stewart diesel engine is a tale of intrigue and deceit, and of intense disappointment. Essentially it is a story of failure. Why the engine failed as a commercial venture is not entirely clear, even today. Maybe the engine just wasn't that good, but certainly World War II, the Great Depression, a lot of bad luck, and an element of bad judgement and greed were factors as well.

In the early 1920s George Edlin (b. 1884, d. 1976) and Hector Halhead ('Steam') Stewart (b. 1888, d. 1950) developed a new kind of engine. It was a horizontally opposed two-stroke engine. Two-stroke engines are inherently efficient as compared to four-stroke engines, simply because it only takes two piston strokes to deliver the power – so there's less friction and less movement for the same power stroke. However, a two-stroke engine has trouble efficiently flushing the exhaust from the chamber and taking in new fuel-air mixture – that's what the other two strokes in the four-stroke are for. This means that, especially at low revs, the two-stroke will always run rough. The Edlin-Stewart engine was more accurate and efficient at getting rid of exhaust and taking in new mixture. Two pistons shared a common cylinder, and a sleeve inside the cylinder moved in opposition to the pistons. Holes in the moving sleeve formed the valves and both the sleeve and pistons were connected to cranks. Tests at Auckland University throughout the 1920s confirmed the engine's power-to-weight advantages, and it was accidentally discovered that it was also a very good diesel engine, despite being designed to run on petrol.

The story of the Edlin-Stewart diesel engine is a tale of intrigue and deceit, and of intense disappointment. Essentially it is a story of failure. Why the engine failed as a commercial venture is not entirely clear, even today.

Diesel was big news during the depression because it was cheaper than petrol. Diesel engines, however, were generally big and heavy so they were not used in cars, only in buses and trucks. If the Edlin-Stewart engine could be developed as a diesel engine for cars, the sky was the limit – well, the road was the limit. In 1930 the *Sydney Morning Herald* had this to say about the invention: 'The new engine is ingenious and yet surprisingly simple; and judging by its reception by English experts, its future is promising. If it lives up to its promise, then possibly it may yet mark a new era in the history of internal-combustion engines, for by its use a four-cylinder engine would give the smooth even torque of an "eight" with approximately only half the weight and size.'

A large public company was formed. Sixty thousand pounds was raised from public subscribers – who might well have been my grandparents or yours! – and Stewart was sent to show the engine off to American car companies, hoping for a bite. Despite the depression having stopped production in car factories all over America, some car companies were still doing research and development. Stewart impressed the Dodge company enough that they offered the Edlin-Stewart company £10,000 – almost $1m in today's money – to road test the engines for nine months. During that nine months, the company would not be allowed to sell the engine to anybody else, but if at the end of the nine months the engine was proven worthy, the Dodge company would pay a royalty to the company for each engine that was built. Dodge built a lot of engines, and if the engine was all it was cracked up to be, that deal could have made the shareholders very wealthy. But the company directors turned it down. Ten thousand pounds wasn't much to tie up their engine for nine months, they reasoned. So far into the future, it's hard to say for sure, but that decision certainly could be the greed and bad judgement part of the story.

And here is the bad luck part. Up until this time diesel engines had been too heavy for aircraft applications, but famous aircraft manufacturer Howard Hughes saw the potential for diesel aircraft engines. If the Edlin-Stewart engine could power aircraft the sky really would be the limit. Hughes arranged to meet Stewart in a New York restaurant. Stewart's taxi driver had other ideas. He took the Kiwi to New York's dock area where Stewart was robbed and left with nothing. He never did have the meeting with Hughes and the company back home decided Stewart should try and peddle the engine in England.

Despite having a wife and six sons in Auckland, Stewart headed for the UK, and what followed for him was three and a half years of fruitless struggle in London. England was different from the US. The American dream stipulates that if you are nobody you have still got a shot. There is not even a thing called the English dream, and Stewart had a hard time getting anyone to listen to him or look at the engine. The Bristol Aircraft Company showed interest, but their head engineer was developing his own engine, so (strangely) nothing happened. Money ran out. Stewart's supporters in London and in Auckland got fed up and wound up the operation. Stewart was left stranded in London. The engines were lost – probably melted down and fashioned into railings, pokers and nutcrackers – and it took Stewart seven months to work his way back to Auckland to his family.

Was the engine ever a real prospect? According to the Stewart family the engine had had a lot of exposure by the time World War II broke out, and patents had been lodged internationally, including a US patent in 1936. These patents, they say, were stolen and incorporated into the world's first successful diesel aircraft engine – the Junkers Jumo 205, made by some thieving Germans. Stewart still held the patents for the Edlin-Stewart engine and was planning a post-war lawsuit against Junkers when he died in 1951. Just to add insult to injury, jet engines came along and made diesel aircraft engines unnecessary.

Maybe you're thinking, 'Hey! War, depression, deceit, greed, ingenuity, fate and luck – this all sounds like a great movie!' Well back off, it's my idea.

The Smitkin Engine

THE LITTLE ENGINE THAT COULDN'T

The engines under the bonnets of our cars are fundamentally the same as those invented in 1876, even before there were cars. Inventors all over the world have assumed there must be something better. For over a century, the search for a better petrol engine has been the inventor's gold rush.

A new engine will need two major advantages to overcome the difficulty of a worldwide shift to a new technology. Firstly, it will need to be more fuel efficient. The engine that we all know is no more than 26–27 per cent efficient: that is, only about a quarter of the energy released by the exploding gases is delivered to the gearbox. It doesn't seem that 27 per cent efficiency should be a difficult mark to surpass, but it is. Secondly, it will need to have zero emissions. Further to those, it will have to be just as light, just as cheap, just as easy to work on…

The Smitkin engine was invented in New Zealand by two Aucklanders, Roger Smith and Graeme Jenkins. After initial testing, it caused a great deal of excitement, both here and overseas. In a press release in March 1997, Jenkins said, 'It will change the world, provided somebody runs with it.'

Smith first came up with the idea in the early 1990s. The engine shared a lot of the benefits of the famous Wankel rotary engine but was far more efficient. Instead of having a crankshaft, the entire Smitkin engine was designed to spin. The engine was lightweight – a prototype Smitkin made of aluminium weighed around 24kg – against a similar two-litre combustion engine of 140kg – and was about the size of a car wheel without the tyre. The diesel Smitkin had 30 parts; a conventional diesel engine has around 3000 parts. The engine contained no crankshaft, there were no cams, no chains, valves, cogs or bolts inside the engine. The ignition system did not spin with the engine, but the spark jumped directly to the spark plug as it passed a wiper at the desired point for the ignition to occur.

It was estimated the engine would be up to 60 per cent cheaper to build. Fuel consumption was 40 per cent better than conventional engines, it was air-cooled and thus required no radiator, and torque was such that vehicles would require fewer gears. The pistons in the Smitkin engine did not change direction as the engine ran, so there was no running vibration. It was easy to manufacture, and easy to assemble and work on. Smitkin's inventors reported a 30 to 40 per cent decrease in emissions – a feat auto manufacturers at the time would have dearly loved to be able to match. It sounds almost too good to be true.

A deal was struck with Chinese firm, the China Sichuan Donghua Machinery Works, to bring the engine to production stage. It seemed to be the invention that would finally unseat the ubiquitous combustion engine.

But it wasn't. During testing, the engine would overheat so much that they blew up several engines. Eventually, the Chinese company dumped the engine and Smith reportedly set to work on yet another engine – this time an efficient, completely emissions-free rotary *external* combustion engine. Smith's words in 1999 were, 'Watch this space.'

The Smitkin engine company went into voluntary liquidation in 2006. Maybe there just isn't a way to make a better engine.

Tretech

COUNTING RINGS IS EASIER THAN COUNTING MONEY

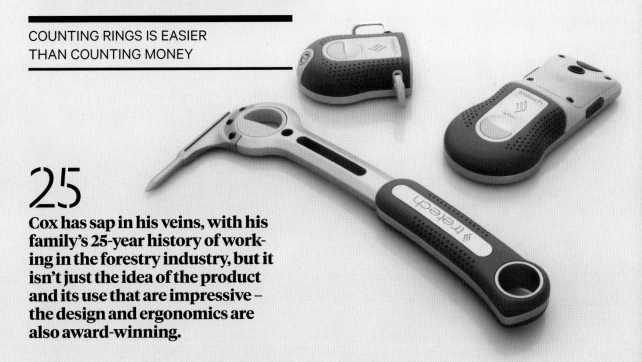

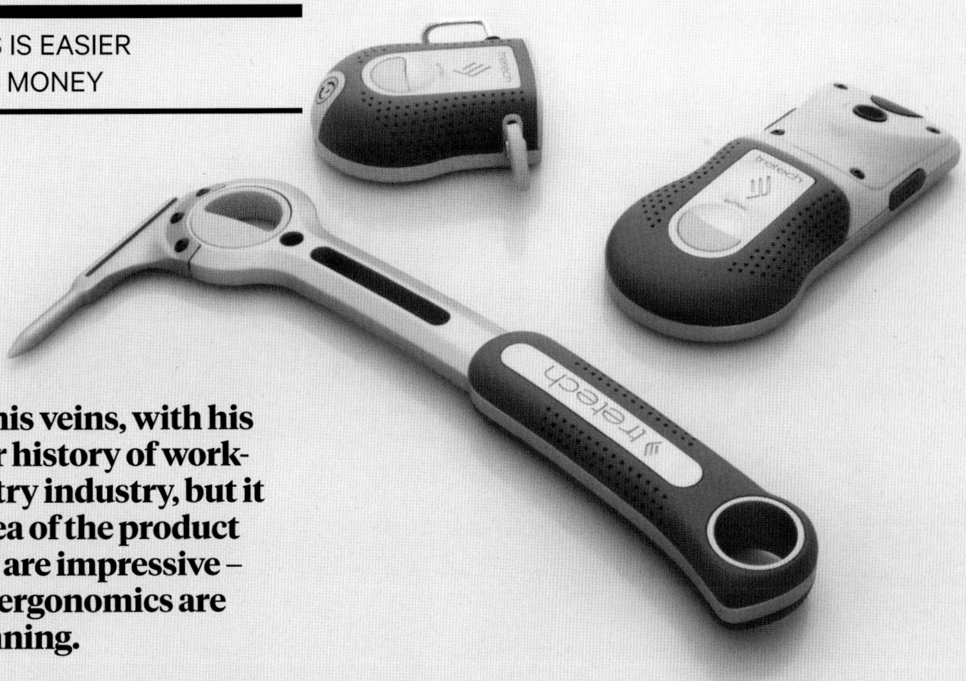

25

Cox has sap in his veins, with his family's 25-year history of working in the forestry industry, but it isn't just the idea of the product and its use that are impressive – the design and ergonomics are also award-winning.

New Zealand has a lot of trees. It's probably even possible to count how many without *too* much difficulty, given enough time and eyeballs. But knowing how much each tree is worth, if it's the right type to cut down for timber, and when is the right time to do that – this is a much harder problem, certainly not one solved by meandering through a pine forest and idly glancing around.

Christchurch designer Tim Cox designed a product to help – his Tretech system, which is really three products in one. A handheld hammer goes into the tree and anchors there, then an ultrasound transponder device measures the diameter, quality, and density of the wood, and sends all that data back to a handheld receiver, which also measures the height of the tree. So now you have all the data you need to assess the quality of the tree. Add to that GPS and a camera and you've got a full system for tree information that replaces older, more expensive products and can eliminate double handling and errors.

Cox has sap in his veins, with his family's 25-year history of working in the forestry industry, but it isn't just the idea of the product and its use that are impressive – the design and ergonomics are also award-winning. Cox entered the Tretech system into the prestigious James Dyson awards, held each year across the globe to find the most innovative and well designed new products by young and up-and-coming designers. Tretech won the 2009 competition and with it, prize money and a trip to the Dyson design facility in the UK.

However, the Tretech story doesn't end there. After getting a few smaller investors on board, Cox started to come up against the classic barrier for start-up businesses – money. The costs required to keep the company going and to keep innovating and developing the products meant that the full dream of an internationally commercial model began to slip away. Cox had to mothball the Tretech as a product – showing that even an award-winning design can sometimes not be enough to ensure success. Like most successful entrepreneurs, though, Cox is determined to learn from the experience, and to come back with more products. The experience of trying to commercialise Tretech will be crucial to the success of his new company, Tinka Design. With this attitude of 'there is no failure, only learning', no doubt Cox will continue to be a design and innovation leader in the future.

Instant Coffee

NO EXPERIENCED COOKS REQUIRED

1889

In 1889, coffee and spice merchant Strang created his Dry Hot-Air process to convert coffee into soluble granules and then set about patenting it in 1890 – one of the first ever applications of New Zealand's new patent laws.

Right: The world's first instant coffee. And chicory powder.
Far right: Today's coffee granules are Strang's legacy.

'Strang's soluble coffee powder requires no boiling, but is made instantly with boiling water. Then, again, it can be made in a breakfast cup and requires neither the use of pots nor the employment of experienced cooks.' So said the *Otago Daily Times* in the late 1800s, heralding the invention of Invercargill local, David Strang (b. 1847, d. 1916). It may read as quaint to the modern eye, but when you consider that our morning coffee *does* require 'the employment of experienced cooks' (only we call them baristas now), you'll realise Strang was on to something – a way to turn a liquid morning staple into a soluble powder that magically reconstitutes with boiling water.

In 1889, coffee and spice merchant Strang created his Dry Hot-Air process to convert coffee into soluble granules and then set about patenting it in 1890 – one of the first ever applications of New Zealand's

However, while the debate about whether instant coffee tastes better than the freshly brewed stuff rages on, there is no debate now as to the original inventor – another example of the deep south of New Zealand's impact on the world.

new patent laws. Strang's Patent Soluble Dry Coffee-powder, as it was called, had the advantage over coffee beans in that it was easier and lighter to ship and had a good shelf life. He packaged it in tins, as he did with his other spice products, and set about marketing and distributing his new invention.

At first Strang's supplied this and other products only locally, but they eventually expanded to national distribution and even internationally to Fiji. The products became very well known, and tins of their spices and other products adorned the shelves of many a pantry in the 'colony', as we were described at the time. It was probably this name brand recognition that helped promote the new instant coffee, as it seems that until this point, the consumption of coffee was not common in New Zealand.

Like many inventions, there was some confusion as to the originator of instant coffee, and for many years a Japanese scientist called Satori Kato, who had been working in the US in 1901, was given the credit for it. Following Strang's invention, others have created new and different ways of making instant coffee, with techniques such as freeze drying and spray drying being the most popular methods today. Indeed, Strang's technique for creating instant coffee – sorry, soluble dry coffee powder – is no longer the favoured approach; other mechanisms have proved more efficient.

At the time, however, Strang's technique for creating coffee was new, and even the coffee snobs of the time were complimentary, with the *Otago Daily Times* saying, 'None of its natural fragrance is destroyed, as is undeniably the case with the vast majority of extracts and essences.' A red ribbon roast, despite what they might say in Ponsonby cafés today.

However, while the debate about whether instant coffee tastes better than the freshly brewed stuff rages on, there is no debate now as to the original inventor – another example of the deep south of New Zealand's impact on the world.

Tullen Snips

THE STORY OF A MAN WHO HAD A VISION OF BETTER SCISSORS AND (DESPITE HIS MOTHER'S ADVICE) RAN WITH IT

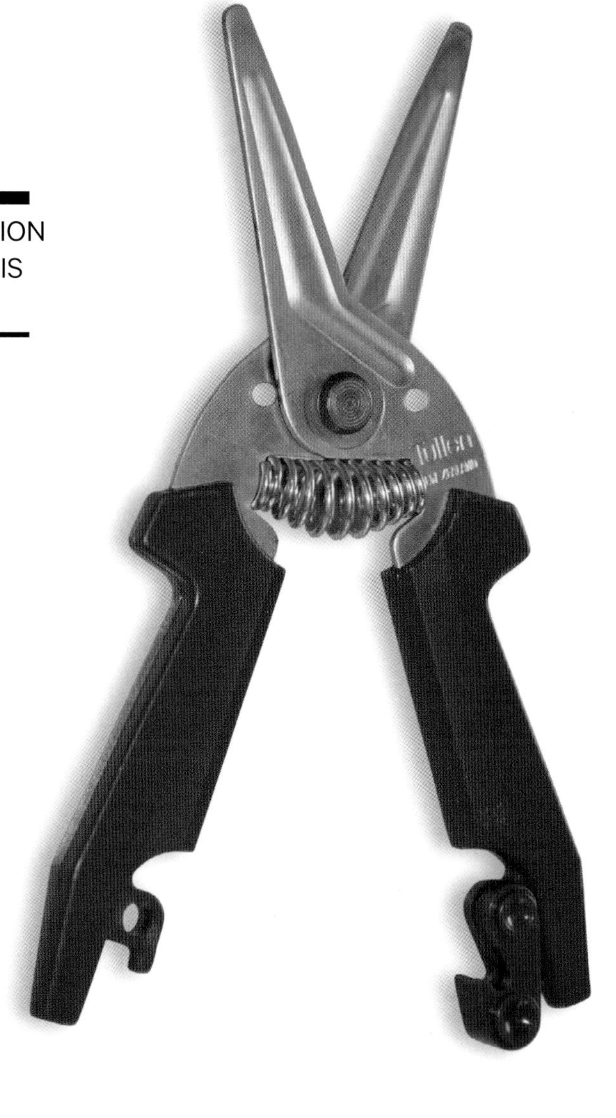

20,000

Try as they might, the process eluded them. They threw out the first 20,000 or so pairs of shears they made.

Right: A pair of hard-working snips take a well-earned rest.

'It is amazing what you can do when you start with no preconceptions from a different angle,' says the inventor of the Tullen Snips, John Hough – to which he could have added, 'and when you don't really know what you are doing'.

Tullen Snips is a product familiar to a generation of New Zealanders. They burst onto the scene in the late 1970s as the scissors that could cut anything, but weren't pointy, so they'd never cut you. While scissors have been around for a long time (in fact pivoted scissors made of iron were used by the ancient Romans) the success of the Tullen Snips is due to an entirely Kiwi accident.

The blades were so tough they could cut anything from paper and plastic wrap to one-cent pieces. They were easily the most versatile and best-cutting home scissors in the world.

In 1972 somebody handed John Hough a pair of shears made overseas, suggesting he duplicate the manufacturing process and make them himself. The advantage they had over normal scissors was

1978

Meanwhile, the Mini Snips – devised for the government to buy as school scissors – were in almost every classroom in the country by 1978.

extreme cutting power. The shears had been manufactured with a heat-treating process which hardened the mild steel blades. Hough set out to copy the heat-treating process, which involves dipping the steel into hot molten salts in order to try and get an even heat throughout. However, his engineers didn't really know what they were doing, and, try as they might, the process eluded them. They threw out the first 20,000 or so pairs of shears they made. The breakthrough came from trying to jam too many blades into the heat-treating vat so they could produce the shears cheaper. They got what could be called a 'poor heat-treating result', but the uneven heating turned out, by pure accident, to make the blades much more resistant to chipping.

Hough sent the resulting shears to England for analysis, and the English engineers described the shears as having resulted from a 'very sophisticated heat-treating process'. Normally the blades of shears had to be heat treated, then cut out, then sharpened with an expensive diamond-head sharpening process. The accidental Kiwi process meant that the steel could be cheaply punched out with a power press, then heat treated and the resulting blades would require no sharpening. Furthermore, the cheap pressing gave a slightly rough edge which meant the blades gripped as they cut, making them even more powerful. The hardness meant the blades could be made to press much more firmly together than normal scissors; and a slight bend in the blade meant they touched at one cutting point that moved up the blade as you squeezed.

Hough went into production with what he dubbed Snips. The blades were so tough they could cut anything from paper and plastic wrap to one-cent pieces. They were easily the most versatile and best-cutting home scissors in the world. The Tullen company underwent a steep learning curve in marketing, exporting, design and patent protection. They hired marketing expert Steve Bridges, who discovered that although the product cut well, its utilitarian design turned consumers off. Bridges' market research, and product design by Peter Davies, led to new products – kitchen, children's and handyman styles – with the same cutting power as the originals. What followed was fantastic commercial success, both in New Zealand and overseas. The wall-holder for the kitchen Snips turned out to be a more easily patented innovation than the Snips themselves, and helped battle the patent infringements of more than 100 copycat products from Taiwan and Korea.

In the days of the Muldoon government's export incentives, Snips sold in huge volumes overseas and became the category leader in Australasia, North America and South Africa. By the time John Hough sold the company and its processes and patents to Wilkinson Sword in 1985, over 20 million Snips had been shipped, with 90 per cent of them exported to 30 countries. The biggest single order was for 3,250,000 Snips to go out as a bonus on dish-washing powder packets in Italy. Meanwhile, the Mini Snips – devised for the government to buy as school scissors – were in almost every classroom in the country by 1978. Show them now to New Zealanders of a certain age and they'll cry, 'I remember those!'

Why are Snips no longer around? It seems Wilkinson Sword were more interested in the processes Tullen had evolved for the production of normal scissors. They broke up the Snips factory and sent it to Bridgend in Wales, where they made Snips for a while – but they never got them quite right. An iconic Kiwi brand and invention faded away.

Hough is still innovating, working on projects that don't involve Snips. Davies says the technology is still there, and a factory could be set up to once more make the best cutters in the world, but it would cost so much he doubts it'll ever happen.

Expanded Polytetra-fluoroethylene

GORE-TEX COULD HAVE BEEN CROPPER-TEX

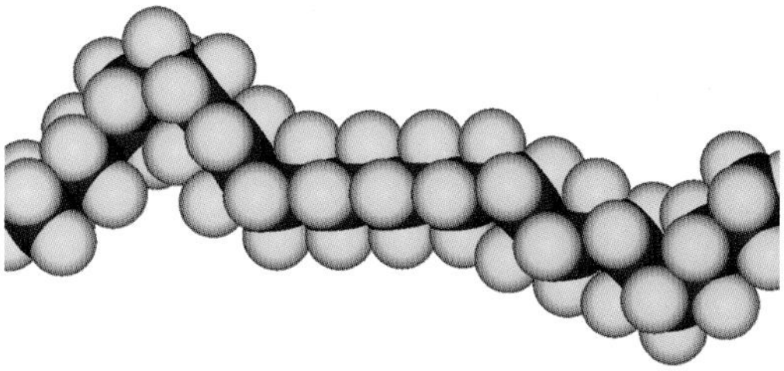

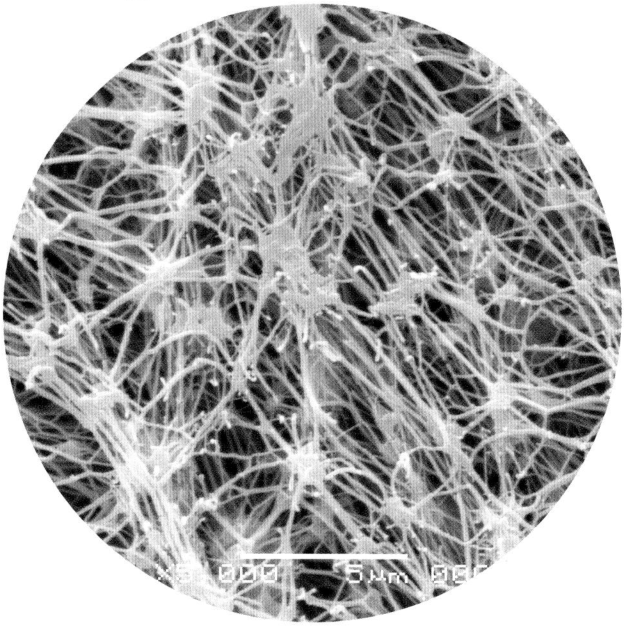

If you search through the wardrobes of your house, you are likely to have some. The waterproof fabric known as Gore-Tex is a technological marvel – it lets sweat and moisture out, but protects against wind and rain. It's smooth to touch, it's a multi-billion dollar industry – and it should have been ours!

This is a sad tale, another near miss and another lesson on protecting intellectual property. It's also a good lesson in how patent law works, and what inventors need to think about before they go public with their ideas.

In 1966, in New Zealand, one John W. Cropper developed and built a machine that could create a material called 'expanded polytetrafluoroethylene' (or ePTFE for short). Cropper was an engineer, building machines to manufacture industrial components, when American chemical giant DuPont approached him and asked if he could build a machine to manufacture Teflon tape. That's just what he did.

Before we go any further, we should explain this stuff. PTFE is essentially a product which we also know as Teflon, but stretched. Teflon, as it is now widely understood, is a brand name for a substance that is very slippery and non-reactive. It coats pots and pans, and often politicians. If you work out how to take PTFE and stretch it, then it will form a tape, the sort of thing plumbers use in tap joints to stop them from leaking. Stretch that tape sideways and it forms an expanded version – like a wide strip of Teflon. It keeps many of its properties but adds some new ones – it's now full of about 3 billion tiny holes per square centimetre, but they are about 1/20,000 the size of a water droplet. So liquid water can't get through, but water vapour molecules, which are much smaller, can. For those who can

You'll note that was 3 years *after* our Cropper had invented his machine to do the same thing – so how come we aren't all wearing Cropper-Tex?

remember fourth-form chemistry, it's a long chain (poly) with four (tetra) atoms of fluorine (fluoro) hanging off two carbon atoms with two hydrogen atoms attached to each (ethylene).

The almost magical properties of this membrane make it sought after and suitable for a number of uses – we all know it from its use in clothing, where the ePTFE is the core of a layered set of fabric which is waterproof yet breathable, fantastic for the outdoors. It's also used in the medical industry – it turns out that human body tissue can grow through it, effectively making it a scaffold for the body – and for protecting old manuscripts.

But its most valuable feature today is probably its name. Gore-Tex is a trademarked brand of W. L. Gore & Associates Inc, whose founder, one Wilbert Gore, came up with a process for creating ePTFE in 1969. You'll note that was three years *after* our Cropper had invented his machine to do the same thing – so how come we aren't all wearing Cropper-Tex?

It's clear and undisputed that Cropper made his machine, and the first ePTFE tape, three years before Gore did. But Cropper didn't patent his machine or his process; instead he decided to try to keep his method a secret. In some ways, that's a valid decision – when an inventor comes up with an idea, they can choose a number of ways to protect it. A patent means that the world gets to find out all about the idea, but that the inventor has a 20-year head start on commercialising its use. A trade secret means that no one gets to find out the secret behind the invention, and the inventor can do whatever they want with it for as long as they can keep it a secret. Coca-Cola is probably the most famous example of a product that remains a trade secret – if you can find out the recipe you are free to make it yourself.

Looking back, Cropper says that if he knew then what he knows now, he would have taken out a patent, but he was under the (mistaken) belief that he couldn't patent a process. So instead, he manufactured the machines and asked customers to keep its process secret. It was a unique machine, producing better quality, softer Teflon tape than anyone thought possible. In fact Cropper believes a UK company attempted industrial espionage – sending someone to work for him to find out all about its manufacture and methods.

In 1968, Cropper sold one of his machines to a US company – the Budd Company. He contractually bound them to secrecy, but the folks at Budd consulted with a third party about how they might make Cropper's machine more effective, and there is some suggestion that secrets were leaked.

Not making the same mistake, in the early 1970s Gore filed for a patent for his process and the product, and it was granted. He set about commercialising and marketing Gore-Tex and it quickly became clear it was going to be a big money spinner. The US Army – and your dad – had been hunting for a lightweight, waterproof, breathable material for years.

In 1979, Gore found out that another company (Garlock) had been using the Cropper-invented machine to churn out an alternative product to Gore-Tex, and sued them for breach of patent. Garlock, using Budd's machine and Cropper's invention, defended the suit, claiming that Cropper was the original inventor of the ePTFE process so they had a right to make it. The US courts originally sided with Budd and Cropper, saying it was clear that the process was first created by Cropper. However, with so much at stake and a legal case to make, Gore appealed the decision, and the US Court of Appeals, after much debate, many expert witnesses and an examination of the facts, overturned the original finding. They essentially found that because Cropper had kept his invention a secret, he had no rights when someone else created their version of the product. In fact, by continuing to create the alternative product, Garlock were now infringing on Gore's patent, and they were forced to stop. Gore was legally recognised as the inventor of ePTFE, and went on to win awards in recognition of 'his invention'. Cropper kept on working, among other things making parts for Fisher & Paykel and inventing a new type of macadamia nutcracker, but he can't help wondering what might have happened if he'd known a little more about patent law at the time.

You might call foul here, and try to make an argument that Cropper was hard done by – perhaps he was, and New Zealand missed out on the multi-billion dollar industry that Gore-Tex became. Fundamentally, though, it's a lesson that while creating the original invention is important, equally important is how you set about commercialising and protecting your idea. Without careful consideration of the right approach, you too might come a Cropper.

11. High Tech

SOFTWARE, HARDWARE, WHITEWARE

New Zealand may portray a clean, green image to the world – and it's great that we do – but we're more than pasture: we are also home to world-leading engineering, science and manufacturing. The tech sector is in fact the third largest export earner in New Zealand (after agriculture and tourism).

These high-tech industries are even more important than we might think. Just because the agriculture sector is our biggest doesn't means we should only spend our R&D dollars there – knowledge gained from R&D in other high-tech areas like those represented in the following inventions commonly spills over into industries, and ultimately feeds the knowledge base of our whole economy. Robotic research benefits our milk-production factories, and electronic petrol pump technology (see page 98) benefits our electric fence industry. The message here is not to treat New Zealand's high-tech industries as novelties, but to prioritise and support them by investing in high-tech R&D, for the sake of the employment and income they offer and for the know-how bounty that they provide. We'll get more bang for our buck.

Inventions today are harder and harder to come up with. The easy ones, the ones a person alone in a shed can come up with, have mostly been taken. Research shows the average age of inventors is increasing, and now it's mostly large teams of highly specialised scientists who invent something that really wakes the world up. That only makes the following inventions all the more remarkable.

Wireless Power

PROF JOHN BOYS AND HIS TEAM CHANGE THE WORLD

75

Today if you buy a TV there is a 75 per cent chance it's made in a factory built from IPT technology licensed to the University of Auckland.

Right: One of the world's largest companies, Qualcomm, has licensed the technology.

1. Power supply
2. Base pad
3. Wireless power & data transfer
4. Vehicle pad
5. On board controller
6. Battery

It starts with a New Zealand professor at the University of Auckland doing a bit of quiet research, and it ends with everyone in the world driving around in electric cars powered by the road itself.

Big, life-changing inventions often come about through a chance discovery. As Distinguished Professor John Boys will tell you, 'If necessity is the mother of invention, then serendipity is the father.' In the 1980s, Boys and his colleague Andrew Green were working on electric power supplies. They made one that didn't do what they wanted it to; but it did supply high power from 'crappy transistors' in a way they'd never seen before. Luck led them to this discovery, but knowing what they were looking at started them investigating inductive power transfer (IPT).

Inductive power transfer means delivering power through the air instead of through wires. In 1831 Michael Faraday discovered induction when he found that a voltage is induced in a conductor when it's exposed to a varying magnetic field. No wires need touch each other: you simply have to create a magnetic field near a coil of wires, then vary that magnetic field and you will induce electricity in the coil. As early as 1894 Nikola Tesla demonstrated a normal bulb being lit wirelessly using induction, but the concept of efficiently and safely transferring real power over a decent distance (up to half a metre) was thought to be impossible.

As Boys continued his research, teaming up with fellow academic Professor Grant Covic in the early 1990s, he began to suspect they could change all that. The time was ripe for a revolution.

There were many problems to solve. Eventually their technology could transfer power wirelessly to deliver high power to multiple loads efficiently. By the mid-1990s they were looking for industry partners. 'We made a model system and we showed it to 26 different companies before one of them said, "Wow, we really like this, we'll go with it."' One potential partner investigated the technology but turned them down on the basis that what they were doing was 'impossible'. Fisher & Paykel offered them a lowly $5000 for it. Finally a company called Daifuku of Japan invested in the technology so they could enter the business of building and selling 'clean room' factories to manufacture electronic goods. IPT technology means that electricity can be delivered to the pristine, dust-free factory floor without an actual connection from the inside of the factory (clean room) to the outside, where the contamination is. On the back of Boys' and Covic's research Daifuku built a US$1 billion business. Today if you buy a TV there is a 75 per cent chance it's made in a factory built from IPT technology licensed to the University of Auckland.

While continuing to develop technology with Daifuku, Covic and Boys and their team of PhD students turned to new applications of IPT – including wirelessly charging electric cars. At its simplest, the idea is that you drive your car over a pad in the ground that charges your car while it's parked. One of the main reasons that people haven't accepted electric cars is the need to plug them in – IPT removes that impediment.

In 2002, they worked with German company Wampfler to set up fleets of wirelessly charged Italian buses in Genoa and Turin – it was the first commercial IPT vehicle-charging the world has ever seen. It was successful, but it was 100 times less efficient than what they are doing today.

There were 'interesting problems to solve': they needed to increase the distance between the car and the pad by 10 times (from 20mm to 200mm) so that cars didn't need to be lowered to charge, they needed 10 times more power for fast charging and 10 times less energy loss. It was these challenges that drove Boys and Covic onwards. One by one they began to solve them.

In 2004 they were ready to partner with a big investor to develop a charging system for electric cars. Despite their proven success, nobody shared their vision. With seed funding from UniServices (the commercialisation arm of the University of Auckland) Covic and Boys set up their company HaloIPT in 2010 and continued to perfect their technology in the face of continued industry ambivalence.

In the end they called in telecommunications giant Qualcomm to help them find an investment partner. To their surprise Qualcomm took a look at what they had and offered to buy the company. Qualcomm is one of the world's biggest telecommunications company, with 26,000 employees and annual revenues of $25 billion. Qualcomm are now developing Covic and Boys' technology under their own Qualcomm Halo brand. Since 2011 they've invested tens of millions of dollars to bring about the dream of wireless charging for electric vehicles on public roads.

Since 2011 they've invested tens of millions of dollars to bring about the dream of wireless charging for electric vehicles on public roads.

The dream of wireless charging isn't just about being too lazy to plug the car in. It means you could get power from the ground where a plug wouldn't be practical – not just in your garage or at work, but while you're parked at the supermarket, while you're stopped at the traffic lights, even while you're driving down the motorway. The big idea is to have inductive power embedded under highways so that the road itself powers cars – the world becomes like a giant slot-car set (without the copper brushes).

If cars are charging little and often like this, then not only will drivers be able to forget they ever need charging, batteries could be smaller so the cost and weight of electric cars can come down – and that will really tip the balance. We could see the end of petrol-powered cars.

Will it happen? Boys says, 'If you'd asked three years ago of the car manufacturers if this technology would be adopted they'd say "not likely". Today it's a no brainer.' And it's all because of a couple of Auckland professors and their students who have been leading the world in this area for 20 years.

And it isn't just through wireless cars they're changing the world – Covic and Boys and their 30 PhD students over the years have created patents now licensed to seven companies, including PowerbyProxi (see page 216). Qualcomm meanwhile keep a team of engineers working in New Zealand and pay both a licence and royalties to the University of Auckland for the technology.

In 2013 Boys and Covic received the Prime Minister's Science Prize – a $500,000 award that they'll use for 'blue sky' research – research that's too 'out-there' for any business partners to invest in: like, say, wireless charging of cars.

Wireless Charging

POWERBYPROXI

1

One of their innovations is what they call loose coupling. This allows a device to sit in any orientation on a charging surface rather than needing to be lined up just perfectly as most wireless chargers do.

Below: PowerbyProxi's charging pad promises to rid the world of cables.

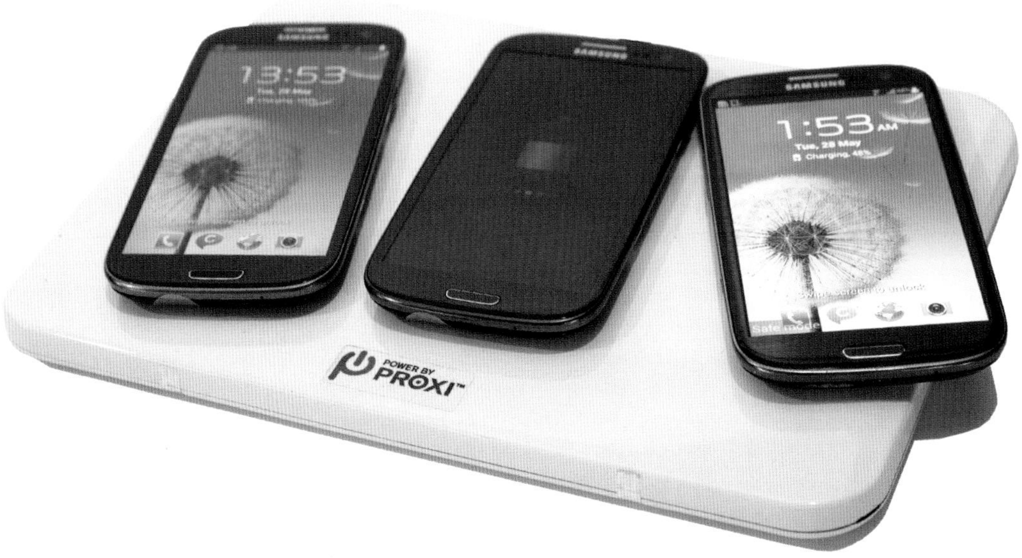

PowerbyProxi is a New Zealand company that has displayed no fear in jumping onto the pointy end of the speeding bullet that is electronics innovation with a series of important inventions in the area of wireless power. While it might be too soon to expect them to be New Zealand's Nokia, they have built on the work of Professor Boys of HaloIPT (see page 214) and now hold an IP portfolio that makes them very attractive to investors in the electronics world.

You might think New Zealand is too small, too distant, or too late to the party to become a player in the gigantic consumer electronics industry, but here it's good to take Finland as an example. When Finnish forestry products company Nokia got into the mobile phone business, it transformed the Finnish economy in less than 20 years. Finland has only 5.5 million people and is further from Asia and Silicon Valley than New Zealand is.

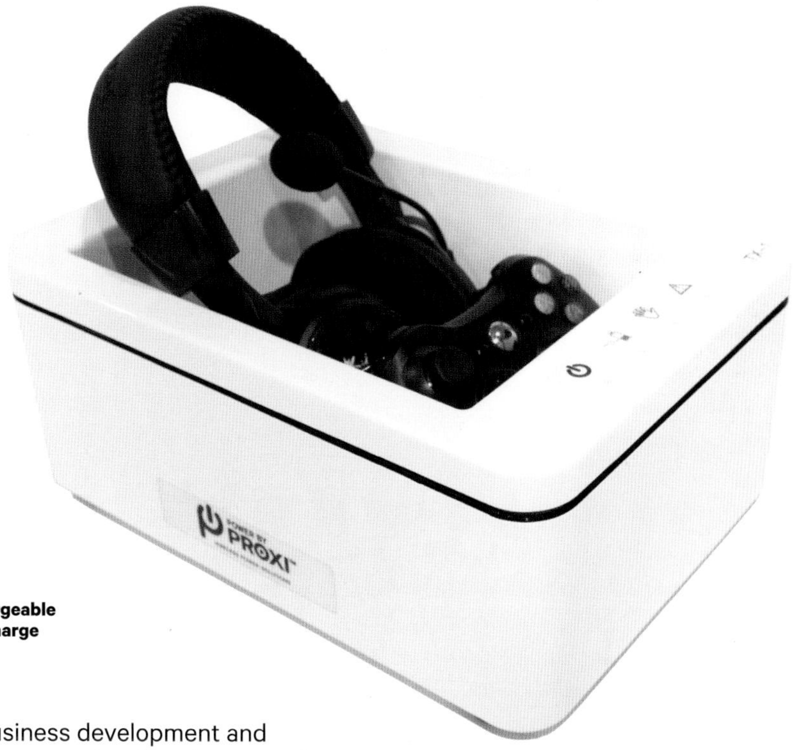

Right: Simply place your rechargeable electronics in a box and they charge automatically.

Through university business development and commercialisation programmes Spark, Icehouse and UniServices, John Boys met Greg Cross. Cross and one of Boys' former students Fadi Mishriki set up PowerbyProxi to license the patents filed by Boys and turn them into marketable products.

That was in 2007. Seven years later, PowerbyProxi are working in both the industrial and consumer space. Industrially they are making wireless power for applications that are too wet, dirty or moving too fast for wires. Their Proxi-Ring replaces a mechanical slip ring and delivers more reliable power with much less maintenance; it has already been used in forestry equipment by John Deere, and in wind turbines by Spain's largest wind-turbine manufacturer.

In the consumer space PowerbyProxi are developing pads and boxes to free people of the necessity of plugging a device into the wall to charge it. Wireless charging will allow them just to place their phone, shaver, tablet, remote control or toy car on a charging surface that could be built into a table or benchtop or to the centre console of your car, and it charges away.

One of their innovations is what they call loose coupling. This allows a device to sit in any orientation on a charging surface rather than needing to be lined up just perfectly as most wireless chargers do. Or, in the case of their 3D loose coupling, the device can be piled up randomly near the surface or in a charging box. PowerbyProxi have also invented the world's first wirelessly rechargeable AA battery. You just dump your device, with the batteries inside, into the PowerbyProxi box and it charges.

Wireless charging will allow them just to place their phone, shaver, tablet, remote control or toy car on a charging surface that could be built into a table or benchtop or to the centre console of your car, and it charges away.

So far PowerbyProxi hold more than 180 patents and patent applications to do with wireless power efficiency, control and miniaturisation, putting them at the crest of the swelling wireless power wave. They have 75 engineers working for them and need even more.

Says Cross, 'It would be fantastic to think we could be to New Zealand what Nokia was to Finland. There is plenty of time for us to be a big winner – a multi-billion dollar company . . . or not. As a company we have to execute. I never ever dreamed I would be part of something as exciting as this.'

If they can pull it off it'll be our greatest inventing story of all: from No. 8 Wire to no wires at all!

High Temperature Super-conductors

BETTER THAN YOUR AVERAGE CONDUCTOR

110

To put the icing on the cake, their superconductor also worked at the positively balmy temperature of 110° Kelvin, or 'only' -163°C.

Right: Flexibility was a key breakthrough for the superconducting wire.

Conductors, in the scientific sense of the word, are things that transport electricity efficiently. The power wires in your walls are made of copper, which is a particularly good conductor of electricity, not losing too much of the electricity in heat as it passes through. Gold and silver are even better, but for obvious reasons your electrician is reluctant to run gold wires through your new extension.

But even these metals provide resistance to electricity, meaning a little bit of energy is lost as the electricity flows through the wires. The more resistance there is, the more costly and less efficient the energy flow is. For domestic energy, this loss of energy can be up to 7 or 8 per cent from the time it leaves the generator plant to the time it powers your telly. On a global or even a national scale, this amounts to a huge amount of energy down the drain.

Superconductors don't have this loss, or at least only lose a very tiny amount of the electricity that passe through them. They promise unparalleled efficiency in powering everything from microchips to power stations to medical scanners. Unfortunately, superconductors only work when it's bollocking cold.

The coldest temperature possible in the universe is -273°C, also known as 0° on the Kelvin scale. At this point matter basically freezes, and all motion of atoms

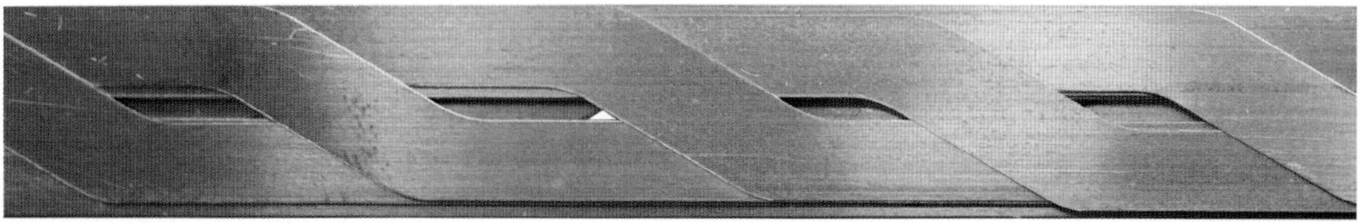

Above: A segment of the future – No. 8 Re-wired.

Working with ceramic superconductors, they isolated a new oxide superconductor made from the elements bismuth, copper, calcium and strontium – a material they dubbed 2223 after the ratios of those elements.

and molecules stops. Not surprisingly, it is extremely hard to create temperatures this cold, but scientists today can get within a half of a degree of it. When cooled to near this coldest limit, certain metals show some remarkable properties – they start to superconduct; that is, to allow electrons to pass through them with no resistance at all.

'Brilliant,' the optimists said, 'within 10 years we'll have 100 per cent efficiency, levitating trains and a better world.' That was in 1911, and so far, few of these things have taken off, apart from the odd levitating train. It turns out that superconductors at these extremely cold temperatures are very expensive to set up. So for many years the search has been on for superconductors that can work at warmer temperatures – that's warmer relative to 0° Kelvin, still not warm enough to be running around in your undies.

Then, in 1988, Wellington researchers Dr Jeff Tallon, Dr Bob Buckley and colleagues at neighbouring IRL made an important discovery. Working with ceramic superconductors, they isolated a new oxide superconductor made from the elements bismuth, copper, calcium and strontium – a material they dubbed 2223 after the ratios of those elements. To put the icing on the cake, their superconductor also worked at the positively balmy temperature of 110° Kelvin, or 'only' -163°C. Buckley remembers as an undergraduate hearing his professor say, 'If anyone can find a way to develop superconductivity at above the boiling point of liquid nitrogen, the world will beat a path to their door.' Tallon and Buckley had done just that.

The discovery of 2223 led, in March 1997, to the creation of the world's first large-scale application of a high temperature superconductor: powering a magnet used for carbon dating. It was created by a New Zealand company, Alphatech International, and it's an ion-beam switching magnet for the particle accelerator at the Institute of Geological and Nuclear Sciences (GNS). The magnet is created by surrounding the superconducting material with liquid nitrogen – at -200°C the gas nitrogen is a liquid. The nitrogen both cools and insulates the superconductor, allowing it to do its job – and unlike other large magnets, if the thing breaks apart no oil or pollutants can leak into the atmosphere; instead the nitrogen 'boils' away as a harmless gas.

The whole area of commercialisation of their work led to the creation of some new companies: HTS-110 Ltd (see if you can work out where the name comes from…) and General Cable Superconductors. They are dedicated to the development and commercialisation of high temperature superconducting magnet systems; their customers are global and they are at the forefront of their fields. HTS-110 sell to industry giants like Pfizer, who use their technology to monitor chemical reactions in their new product development. They also design and sell equipment for nuclear magnetic resonance (NMR) and magnetic resonance imaging (MRI) – and, at last, coils for magnetic levitation. After years of product development, General Cable Superconductors have solved the problem of how to turn a ceramic superconductor into a wire – Buckley likens it to pulling a cup and saucer into a long, thin wire – and their cable is now being used in the world's first power generator to use superconductivity, to be installed in New Zealand.

And all from little old Wellington, where two scientists showed how pure science research could lead to major breakthroughs and significant new global opportunities.

Gudgeon Pro 4 in 1

SMALL BOY TAKES FIELDAYS BY STORM

Right: Patrick Roskam with his patented Gudgeon Pro 4 in 1.

Given that farming inventions usually come from the farm, it isn't surprising that Patrick Roskam of Matamata invented a tool to help farmers accurately hang gates, took it to Fieldays in 2013 where he won a prize and the admiration of Sir William Gallagher, then won the People's Choice Award at the 2013 New Zealand Innovators Awards. It's admirable but not surprising that at the time of writing Roskam is progressing towards production, having just signed off the dies for the plastic extrusion. What is a bit surprising is that Patrick is only 12.

Patrick's father had just bought a farm and was frustrated at having to hang 20 gates from their gudgeons – the metal pins that farm gates hang from. Patrick needed a project for his school science fair and this was a perfect opportunity. The Gudgeon Pro 4 in 1 looks like a 1.4m-long spirit level, but has holes specially placed to guide the drill to allow perfect gudgeon placement. Its length is designed to show you how far to ram the fence posts in; it has markings to guide wire placements; and it has levels to ensure the posts are upright.

12

It's admirable but not surprising that at the time of writing Roskam is progressing towards production, having just signed off the dies for the plastic extrusion. What is a bit surprising is that Patrick is only 12.

There's a long way to go before you could declare this invention a commercial success but Roskam has started his journey in invention early, giving him plenty of time to get there. If you'd like to know how it's going, it's no surprise that you can follow the Gudgeon Pro on Facebook.

Glow-in-the-dark Paths

TRIAL AND ERROR LEADS TO A BRIGHT FUTURE

5.5

Their website has had 5.5 million hits, the product has been on TV in many countries, and enquiries have flooded in from around the planet.

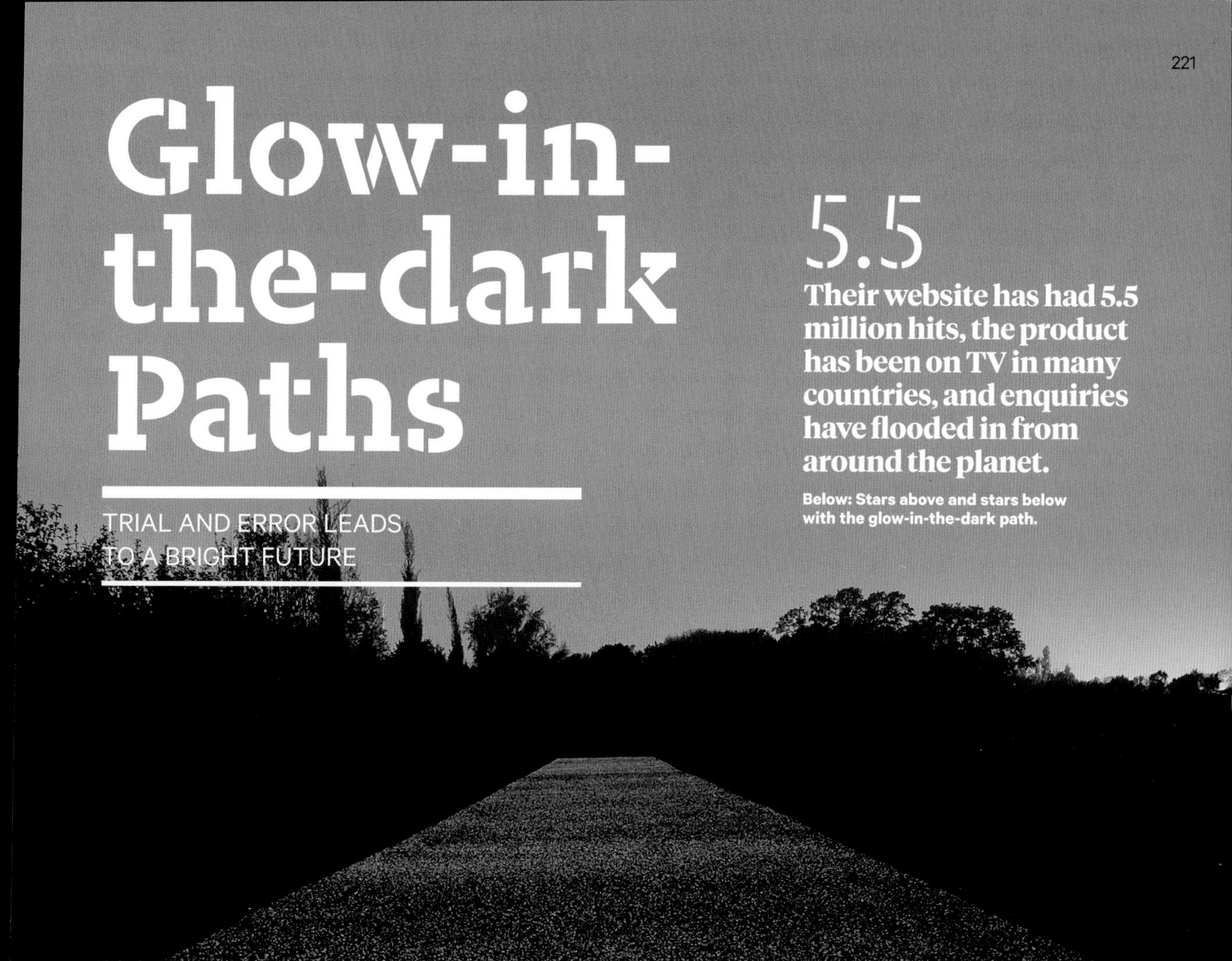

Below: Stars above and stars below with the glow-in-the-dark path.

'It's just another Kiwi invention,' said Hamish Scott, originally from Otaki, when asked about Starpath – his invention that's attracting interest from around the world.

The idea is simple. Where Scott lived in Surrey in the UK, councils had begun to save money by switching off the lighting along paths in parks at night. 'What if the path itself glowed?' Scott wondered. No expensive lighting, no ongoing maintenance, no electricity and the paths could still easily be followed at night.

Scott and his company, Pro-Teq Surfacing, developed and patented a simple solution: a glow-in-the-dark spray that can be applied to a normal pathway. The spray contains a fine aggregate, the particles of which absorb ultraviolet light during the day and glow at night. At the same time Scott's company also developed breakthrough technology which means the sealant also reseals the pathway, repels chewing gum and graffiti, can be coloured, and lasts 25 years.

The development involved pure trial and error over a period of four years, and took place in Scott's shed. 'I'm too thick to understand what I can't do, so I just do it,' says Scott, whose shed is full of things that didn't work. 'I would work on things during the day, then be out there at 4am wondering why it wouldn't glow.'

Their glowing paths are a world first, and the whole thing is so simple there isn't much more to say. Since they told the world about their invention the media attention has surprised Scott and his team. 'We've been bombarded. You put a few stars in a pathway and everybody goes crazy?' Their website has had 5.5 million hits, the product has been on TV in many countries, and enquiries have flooded in from around the planet. Scott says the firm's had a 'massive breakthrough' in the technology recently that could prove even more revolutionary. The old saying is, 'Build a better mousetrap and the world will beat a path to your door.' Scott hasn't built a mousetrap but the world is beating a path, a glowing path, to his door.

Carbon Nanotubes

CLEVER KIWI TURNS DEEP SCIENCE INTO GLOBAL OPPORTUNITY

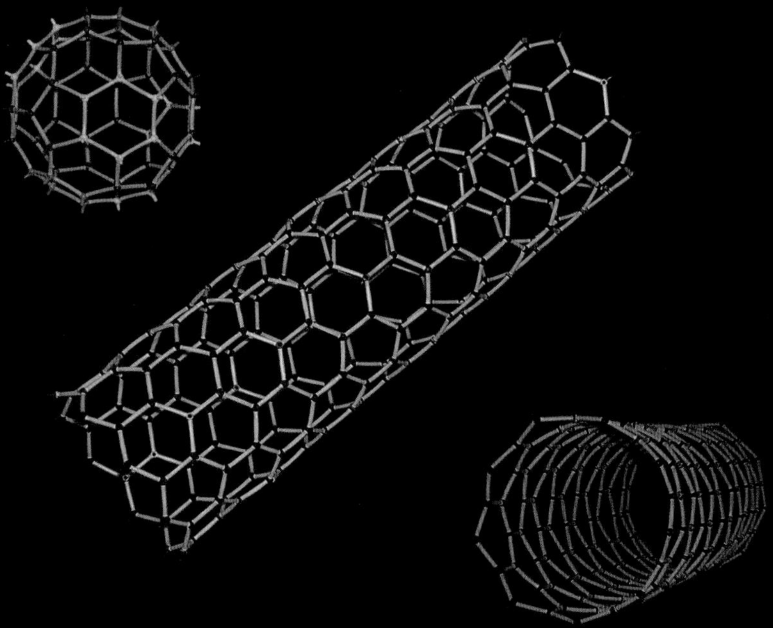

100

Carbon can be things as diverse as graphite (the 'lead' in a pencil), diamond (100 per cent carbon), and even the exotically named football-shaped form Buckminster Fullerene.

Right: Various forms of carbon molecule.

In 1952, at the height of the Cold War, the Russians were onto something – two Russian scientists published a photo in a Soviet science journal that clearly showed extremely small tubes made from carbon. However, due to conditions at the time, and the fact that the article was only in Russian, the rest of the scientific world didn't notice, and the finding fizzled out.

Fast forward to Christchurch, in the 1970s, where scientist Professor John Abrahamson was fiddling around with carbon and electricity, pushing arcs of electrical energy into rods of negatively charged carbon – you know, normal stuff. He was surprised at the end of the experiment to see small fibres left on the rods. They were unexpected and 'jumped out of the electron-micrograph at us, like a forest of very fine fibres or "carbon grass" as we called them.' Unlike the Russians, Abrahamson shared samples of his findings with five of the most well-known labs around the world. No one got that excited – it was too early to recognise their potential. But it turned out that it wasn't carbon grass at all; it was a new form of carbon, where the atoms had arranged themselves into a tube. Abrahamson is now among those credited with discovering this new form of carbon – carbon nanotubes (CNT).

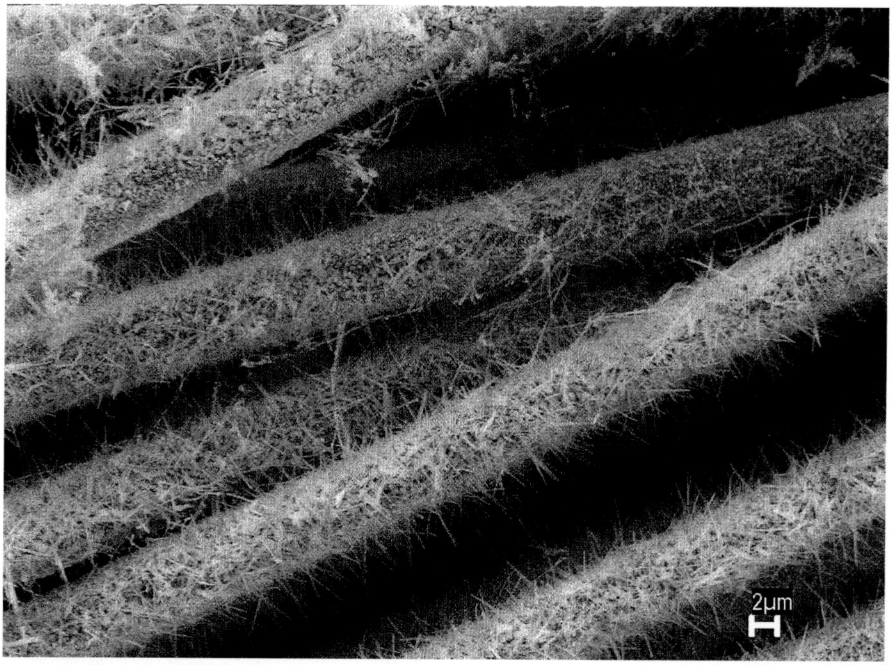

Above: The 'grass' Abrahamson recognised as a new form of carbon.

Carbon nanotubes are a particular configuration where the atoms create either a long tube or tubes nesting within each other. They are proving more than a curiosity, with applications in energy, medicines, the environment and electronics.

Carbon is the basis of all life on Earth – because of the way it has four free electrons ready to bond. I am held together by it, and so are you. It's a basic element with some unique properties, one of which is the way it can form different shapes and arrangements of atoms – meaning carbon can be things as diverse as graphite (the 'lead' in a pencil), diamond (100 per cent carbon), and even the exotically named football-shaped form Buckminster Fullerene (not a joke – look it up). Terms like 'carbon fibre' have made the public more aware that this simple element has some amazing uses, from road bikes to guitar picks to aeroplanes. Graphene sheets are currently very popular as a form. They allow such incredibly thin layers that they're even being contemplated for use in the manufacture of condoms...

But to get us back on track – carbon nanotubes are a particular configuration where the atoms create either a long tube or tubes nesting within each other. They are proving more than a curiosity; with applications in energy, medicines, the environment and electronics. However, that wasn't always the case. After Abramson's description of this new form of carbon in 1979, he expected to see a lot of activity in finding commercial applications, but nothing much happened. In part this was due to the nanotubes being quite difficult to create. So in 2001 he decided to develop a process to manufacture nanotubes efficiently and reliably. The technology he developed is at the core of his Christchurch company ArcActive.

ArcActive is applying the power of carbon nanotube technology to a series of industries. Meanwhile the company are also using their carbon know-how to revolutionise the manufacture of car battery electrodes. It may seem an odd choice but increasingly the market is demanding more energy-efficient cars, and that includes batteries. The potential for the ArcActive electrodes is huge and the interest is high.

As chief technology officer, Abrahamson still steers the science behind the company, and his original arc reactor is about to go on show at Science Alive! science centre in Christchurch. Turns out that it was lucky he didn't get off the 'carbon grass'.

High-tech Innovators

THE FINE LINE BETWEEN INVENTION AND INNOVATION

In researching this book, we found a handful of New Zealand companies that lead the world, but upon close inspection it didn't always become clear exactly what you could point to as an invention. Often these were the companies that came to people's minds when you ask them to think of great New Zealand inventions. In all the cases below the companies have grown to lead the world through ingenuity and innovation. All hold many patents and are at the crest of the technological wave. All have brought, and continue to bring, new things to the world. We couldn't leave them out.

Fusion Electronics

When the brains who created New Zealand company Navman sold the company and went looking for a new challenge they spied car audio company Fusion. Rather than being at the forefront of technology, Fusion was at the 'funfront' of branding, wooing the boy racer audience with Jonah Lomu, hot chicks and a green alien mascot. And they were doing well.

Sir Peter Maire, Chris Baird and the team of engineers from Navman added a new focus: marine audio. They chose this market because the big electronics companies were simply 'painting their car audio products white and calling it marine'. Fusion Marine hit the market with a major innovation. They became the first company in the world to make a marine stereo that you could put your iPod inside. The stereo protects and charges your iPod, allowing you to access the music with a big screen and large knob.

Their innovation won numerous awards, allowing Fusion to take on the big electronics giants – and they have. The company is now the number one marine audio company in the world, growing annually at 30 per cent, with a range of over 80 products. Their Fusionlink software is embedded in the devices of all other major marine electronics suppliers, making Fusion the audio standard. In 2013 they reported over $30 million in annual revenue, most of that from the marine products.

Glidepath

Sir Ken Stevens founded Glidepath way back in 1972 when he took over a five-person engineering firm at the dawn of affordable air travel. Today they are fourth in the world in baggage handling. Glidepath now manufacture in Dallas and Auckland, and have installed 665 systems in 67 countries – many of them costing several millions of dollars.

'There's nothing earth-shattering about what we do here,' says Sir Ken Stevens. Like a lot of products, it's a lot of little innovations combined with an appreciation of the art of mechanical engineering, good software development, and an ability to combine the multiplicity of disciplines needed for large installations like baggage handling. The big three in the area are Dutch, German and American. Sir Ken says Glidepath enjoys being, as the Texans say, 'a burr on their butt'.

When you pluck your bag from a carousel in Dubai, Toronto or Venezuela you can take pride that as a Kiwi, you've sort of delivered your own bag. But also, if your bag goes missing, you've got no one to blame but yourself.

Tait Electronics

The history of radio communications is quite humorous. Mobile two-way radio sets were pioneered in 1923 in Australia by the Victoria Police. They were designed to take the place of the previous method police used to communicate with their base: they stopped and used pay phones. Those first two-way radios took up the whole back seat of the patrol car. Presumably if a criminal was apprehended the cops would say, 'We'll meet you back at the station.'

In 1968 Kiwi tech godfather Sir Angus Tait (b. 1919, d. 2007) set up Tait Electronics and began to pioneer solid-state VHF RTs (or radio telephones) that would fit into the car radio slot. The Tait Miniphone was immediately superior to foreign products available here. They began exporting and an empire was born.

Tait's greatest innovation was probably their world-leading charge into 'trunking standards'. In the 1980s trunking radio systems (where the base station assigns individual calls to available channels) were new and there were no standard protocols to ensure that different manufacturers' radios could work on each other's systems. When the British Ministry of Post and Telecommunications developed a standard called MPT1327, Tait were the first to manufacture systems to work with it. Their first trunked system was for VIP communications at the 1990 Commonwealth Games. Since then they have installed over 400 major trunking systems worldwide, including the well known Fleetlink system that enables all truckies to communicate across New Zealand, and a system for a Colombian mining company which has 18,000 users on it.

Today Tait remains one of the big brands in radio telephone communication in the world. With over 900 employees globally and an annual revenue of over US$200m, they continue to innovate with radio products for large utility companies and emergency services from their Christchurch base. They are currently developing a 'hi-tech campus' as part of the Christchurch rebuild, where like-minded companies can collaborate to fulfil Sir Angus's dream of building an industry, not just a business.

McLaren

The achievements that made Bruce McLaren one of New Zealand's top 10 legends are well known. Engulfed in the sport of motor racing he went to Europe in 1958 and began to drive. His exploits behind the wheel would have made him famous enough – in 1959 he became the youngest ever Formula 1 winner – but it was the team he built that remains his legacy.

From 1966 until today, McLaren is the most successful team in world motorsport. With drivers like Prost, Senna, Lauda, Fittipaldi, Hulme and McLaren himself, it has dominated in F1 and CanAm racing. The success of the team stems from Bruce McLaren's engineering genius and innovative know-how in the early days. McLaren cars became the fastest in the world with his pioneering work in aerodynamics and downforce, and they have stayed that way to the present day.

Though based in Britain, Kiwi DNA and McLaren's inventive spirit lie at the heart of the McLaren Group – a multi-billion dollar company employing 5300 people. In 2013 McLaren released their P1 production car, a hybrid that some are calling the fastest production car in history; Jeremy Clarkson called it 'a genuinely new chapter in the history of motoring'.

Glaxo

'Today Bunnythorpe, tomorrow the World!' Glaxo is the largest pharmaceuticals company in the world. When Glaxo joined Wellcome, the biggest cheque in the history of banking was written. When Glaxo joined SmithKline in the year 2000, it created world headlines and once more established Glaxo at the top. Glaxo started as a milk-drying business in Bunnythorpe, near Palmerston North, in 1920.

In 1904 the idea of drying milk was new. Wellington business icon Joseph Nathan assigned one of his sons the job of setting up a factory in Bunnythorpe to make the brand-new product of the time – milk powder. After some hiccups, including a rival dairy owner burning down the first two factories, the Nathans registered a brand name for their new dried milk in 1907. They called it Glaxo.

Glaxo didn't take off straight away. With the slogan 'Builds Bonnie Babies' the product was marketed in England and in New Zealand as a milk food for infants. Then World War I created a huge demand for dried milk to supply soldiers. The Glaxo arm of the company became bigger than its parent company, and with the addition of vitamin D to their milk products, in 1936 Glaxo New Zealand (and then London) began to branch out into pharmaceuticals.

Paxil, Ventolin, Zantak, Farex, Complan, Beconase, Becotide, Zovirax and Flixotide, Ribena, McLeans and Sensodyne are among the many familiar products manufactured and marketed by Glaxo. As the company grew, the worldwide headquarters shifted to London. Through expansion, acquisitions and mergers it became Glaxo Wellcome, then GlaxoSmithKline. Today, GlaxoSmithKline New Zealand is a small part of GlaxoSmithKline – one of the largest multinationals in the world with $50 billion annual revenue and 100,000 employees, selling over 4 billion packets of products each year to 180 countries. Probably a bit big for Bunnythorpe.

Two-drawer Dishwasher

WHY HAVE ONE DISHWASHER
WHEN YOU CAN HAVE TWO?

2

The two drawers have different uses: a smaller drawer is designed for fast washing of small loads – glassware, cups and mugs – and a larger drawer for utensils, pots and pans...

Right: Teenage readers may not have seen this before – a well-stacked dishwasher.

I think we'd all agree that Fisher & Paykel is a household name. This large manufacturer of whiteware has been part of New Zealanders' lives since it first formed in 1934, although it wasn't until 1960 that they started manufacturing their own designs. Even after their sale to Chinese brand Haier in 2012, they remain a Kiwi icon. F&P have created several innovations – in total over 400 patents – including the SmartDrive™, a clever electronic motor with programmable software built in. This motor has formed part of a number of F&P's products, and led to them creating the two-drawer dishwasher.

It's a simple extension to the standard dishwasher, but on reflection offers a number of advantages – one can separate dishes, only put one on if that's all that's needed, and the two drawers can be arranged nicely in your designer kitchen. Apparently the technology is already popular with the elderly in the US, who, it seems, are reluctant to start the dishwasher when it's only half full.

The two drawers have different uses: a smaller drawer is designed for fast washing of small loads – glassware, cups and mugs – and a larger drawer for utensils, pots and pans, that sort of thing. No longer do fine crystal glasses have to slum it with the filthy lasagne-covered oven dish; in this new class system of cleanliness they can have their own separate abode. Who knows how far this segregation may go? Perhaps next on the horizon is the two-part clothes washer – a separation of whites and coloureds in some kind of clothing apartheid?

The two-drawer dishwasher created a whole new market segment. It was a hit with the Jewish community, whose kosher rules require them to separate meat and dairy products – and the utensils used to prepare and serve them. This became the genesis of a number of other products utilising similar ideas, such as the CoolDrawer, a multi-temperature drawer-based fridge.

F&P are a good example of just how far a good idea can take you. The Chinese obviously liked what they saw, paying over 70 per cent more than the value of the stock price when they took over F&P in 2012, and choosing to leave large parts of the R&D, and even manufacturing, in New Zealand. It just goes to show how much Kiwi ingenuity is worth – and how many dirty dishes there are in the world.

OneBeep

TUNE INTO A GOOD BOOK

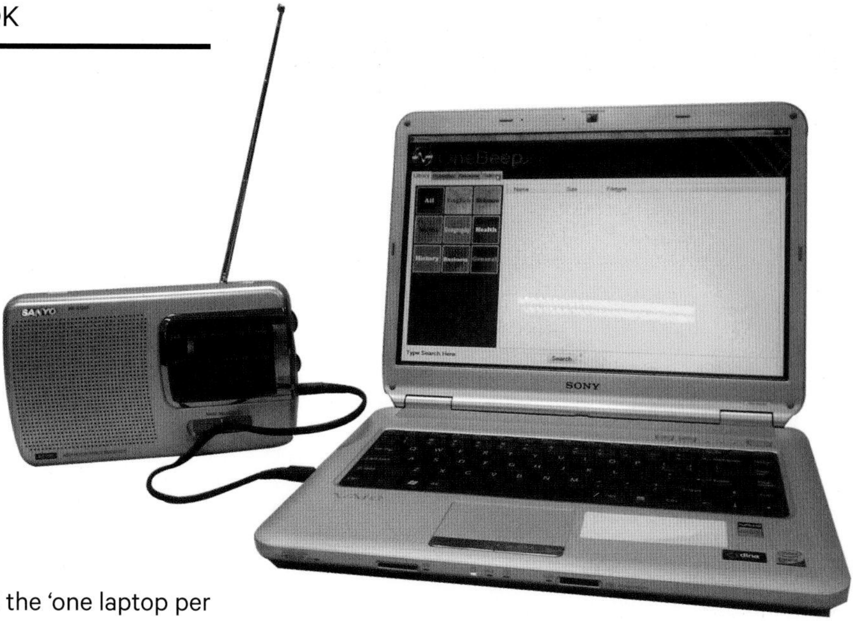

All over the developing world the 'one laptop per child' initiative is providing the world's poorest kids with laptops to help with their education. It's a scheme with one flaw: a lot of those remote areas have no internet, so the laptops become little better than bricks – that you plug in.

In 2010 a group of Auckland students led by Vinny Lohan came up with a nifty solution: transmit digital material to remote communities by ordinary FM radio. Radio waves can travel huge distances and every village has radios. The technology, called OneBeep, works like this: Anybody with the OneBeep software can convert a file like a pdf or an ebook into a normal mp3 sound file. Yes, a book can become a song! To a computer it's all just 1s and 0s, but the mp3 file can be broadcast by an FM radio station. You can tell everyone, 'Maths book being broadcast on 96.3FM at 6pm on Wednesday.' Then the villagers just plug their radio headphone jacks into their laptop microphone jacks and the OneBeep software in their laptop converts the mp3 'song' file from the broadcast back into an ebook and bingo, remote internet!

What followed was a story of huge success and then great disappointment. In 2010 the OneBeep team took their idea to Microsoft's Imagine Cup competition in Poland and out of 400,000 students from 190 countries, placed third. In 2011 they won

2010

In 2010 a group of Auckland students led by Vinny Lohan came up with a nifty solution: transmit digital material to remote communities by ordinary FM radio.

Above: Transmitting data via public radio could transform the developing world.

Auckland University's Spark Challenge and in 2012 entered the Icehouse business incubator. But it soon became clear that to commercialise the idea worldwide was going to be expensive and difficult to do from New Zealand. Further funds became elusive and the team drifted apart.

Vinny Lohan moved to India and at time of writing was determined to launch a Kickstarter campaign to raise the $200,000 needed to release the free, open-source software allowing communities to unlock the power of OneBeep. 'I won't let it die,' says Lohan.

Nuclear Magnetic Resonance

SIR PAUL AND THE PLACE WHERE TALENT WANTS TO LIVE

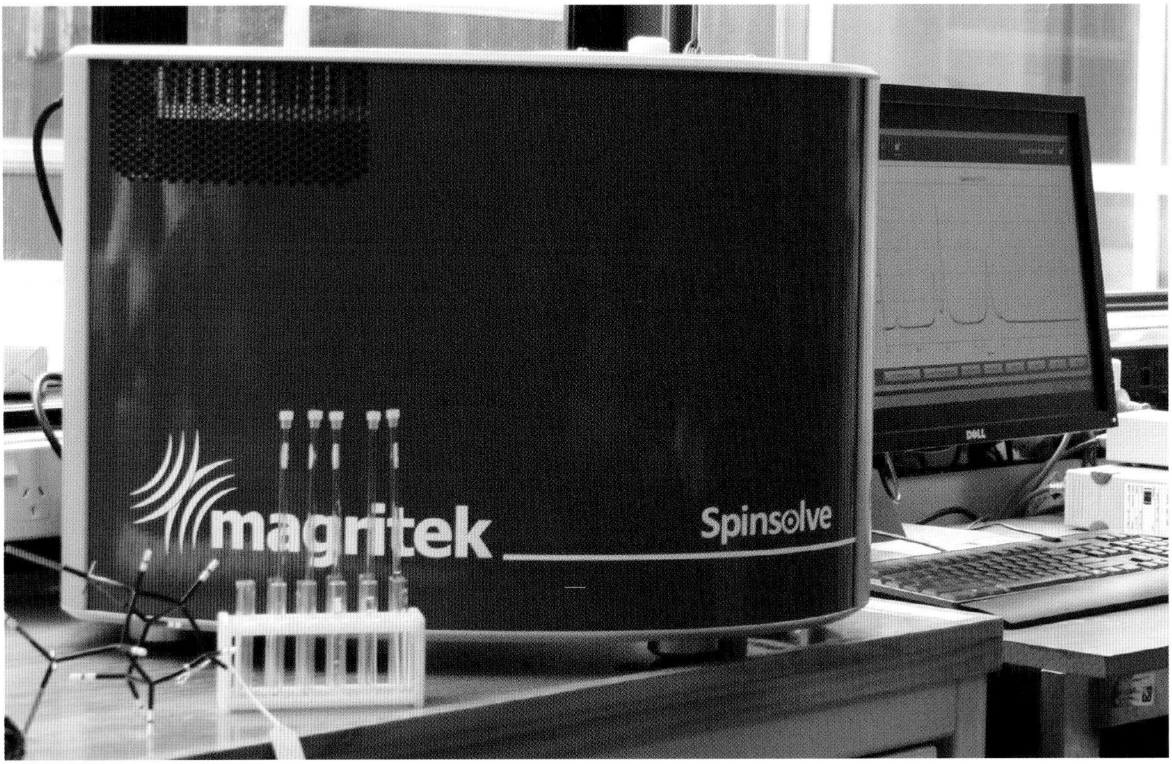

When Sir Paul Terence Callaghan GNZM FRS FRSNZ (b. 1947, d. 2012) died, New Zealand lost more than a scientist; we lost a visionary thinker, with a dream for our country that deserves to be constantly revisited. Callaghan saw a future for New Zealand that encompassed our past but also harnessed the real potential of the country – the people. Callaghan thought, wrote and spoke often about New Zealand's innovation economy. He believed that investment in the dual imperatives of the innovation system of New Zealand, and the quality of life we enjoy, could mean a more prosperous future. His recommendation to create 'a place where talent wants to live' is one to which we can all aspire.

Left: Whanganui's finest, Sir Paul Callaghan, with Prime Minister John Key. Far left: Magritek's devices are built on the science championed by Callaghan.

His natural charm and skill as a speaker saw him sought after for his opinions and he wrote a number of books and opinion pieces on New Zealand's place in the world, setting in progress a new conversation about innovation and the New Zealand psyche.

Callaghan was a New Zealand physicist whose area of study and research was nuclear magnetic resonance (NMR). This is a way of tracking molecules using radio waves. Most people are aware of the power of large MRI machines in healthcare, where signals from the magnetic fields interacting with the atoms inside the body can be used to map the inner structure of the body. Callaghan pioneered techniques using magnetic field gradients and achieved a number of world firsts, including the imaging of the internal structure of a microporous material and the observation of the flow profile of a complex polymeric liquid during shear banding.

While that sounds (and is) difficult to understand, the techniques he developed allowed for the science to explore more varied uses, for example, to look at oil production, plastics production and even food products. Dairy giant Fonterra, working with University of Auckland UniServices, recently used his techniques to develop an improved form of mozzarella cheese.

Honouring another great New Zealand scientist, Callaghan went on to set up a centre of research excellence called the MacDiarmid Institute for Advanced Materials and Nanotechnology (see page 194). The institute spans across a number of labs and universities in New Zealand; the host is Victoria University in Wellington. It conducts research in six main areas: nanofabrication and devices, electronic and optical materials, molecular materials, soft materials, inorganic hybrid materials, and the intersection of nanoscience and biology.

With the MacDiarmid Institute, Callaghan showed his skill outside the laboratory, creating a world-leading collaboration of Kiwi scientists. His natural charm and skill as a speaker meant he was sought after for his opinions. He wrote a number of books and opinion pieces on New Zealand's place in the world, setting in progress a new conversation about innovation and the New Zealand psyche. When we look back, Callaghan will be seen as a pivotal figure in New Zealand's history.

He was not only an academic; his work led him to set up a new company, Magritek. Today a world leader in MRI instruments with millions of dollars of export earnings, Magritek produces portable NMR machines used for teaching and as analytic tools.

Callaghan was a giant in his field, and he collaborated and taught widely. A recipient of many awards, he was a proud Kiwi. Born in Whanganui, he spent most of his life researching and working in New Zealand, and served as president of the International Society of Magnetic Resonance. Like MacDiarmid before him, Callaghan's contribution to science and innovation in New Zealand is recognised, with his name being applied to the new government agency Innovation. Sir Paul Callaghan has been described as 'the greatest scientist to ever ply their trade in New Zealand'.

Seismic Isolators

SHOCK ABSORBERS FOR BUILDINGS

Right: The LRB isolators being installed and tested.

New Zealand has always taken the threat of earthquakes seriously. Suffering from a succession of earthquakes that have punctuated our short history with tragedy, we pride ourselves on our world-class earthquake readiness. Despite that fact, earthquakes still spring very nasty surprises on us, so the pioneering research undertaken by our seismic scientists has made a great difference.

Nobody is owed a greater debt of gratitude around the world, as well as in New Zealand, than Dr Bill Robinson (b. 1938, d. 2011). Robinson grew up in West Auckland, and studied engineering at the University of Auckland before completing a PhD in physical metallurgy at the University of Illinois. He then brought his knowledge and nous back to New Zealand where he worked for the DSIR (now Callaghan Innovation) from 1967 to 1991.

One day in April 1970 he joined a morning tea discussion with Dr Ivan Skinner at the DSIR offices in Gracefield, Wellington. Skinner was the head of engineering seismology at the DSIR and had recently returned from a sabbatical in Japan. Skinner told Robinson about his plans to use steel earthquake dampers in the base of the new government building being planned for the Ministry of Works – the William Clayton Building.

Robinson went back to his office to think about the problem. His metallurgist's instinct told him there should be a better material for damping than steel. The material needed to be face-centre cubic, which means that the atoms are arranged in a cubic crystalline structure within the metal. This gives shear planes in three dimensions and so allows the bearing to respond to force from any direction. But the material also needed to have a low melting point so that after it is deformed by a quake, it can come back to shape and the crystalline structure of the metal can re-form perfectly. Robinson went back to the primary source of information for a scientist, the periodic table, and after two hours came up with the answer: lead!

Robinson and his team went to work designing what is now known as the lead-rubber bearing base isolator or LRB. There were a lot of difficulties to overcome, and the process took several years; the lead bearings they made failed too easily. It wasn't until 1975 that Robinson came up with the solution – to surround a solid lead core with a supporting structure of laminated layers of steel and rubber. Two more years of work and they were finally ready to test the method. They bought an old Caterpillar tractor to generate the force of an earthquake, and after some trial and error they hit upon the right design.

Robinson's LRBs are installed between the building and the foundation – like piles to hold the building up. For most of the time, they just sit there, supporting the building with little or no fanfare. Virtually forgotten, they silently guard the occupants against the tyranny of earth movement. When an earthquake hits they spring into action.

The ground beneath the building moves from side to side, and given the tremendous weight of the building above, the top can't move as quickly. The building is under immense pressure to shear – physics-speak for 'tear itself apart'. Furthermore, as the earthquake progresses the building actually magnifies the earthquake so that the top of the building ends up wobbling from side to side more than the base, which further increases the shear forces within the structure.

With LRBs in place, the sheer shear pressure of the building above compresses down on one side of the shock absorber, and a weird thing happens. The huge forces at play cause the lead at the core of the LRB to melt for an instant, taking the pressure of the building above and the ground below. It 'gives'. Then, as the shock passes, the rubber and steel rings snap the lead back into shape, where it re-forms an instant later. The isolator passes only a fraction of the movement from the ground into the building.

In physics terms the lead in the bearing undergoes what's called a plastic transformation – ironically it's a bit like the transformation we call liquefaction, which occurs in sandy soils under earthquake pressures. This phenomenon is fairly uncommon, but luckily lead displays exactly the behaviour needed by Robinson and his team.

Finally, in 1978 the technology was ready. The first building to have the LRB installed was the William Clayton Building – the government building they'd been thinking of over that cup of tea.

The use of Dr Robinson's invention spread quite slowly at first: it was installed in a few buildings in earthquake-prone parts of the world. When first Northridge, California, then Kobe, Japan were struck by earthquakes in 1994 and '95, the isolators were faced with their first real-life tests.

During the 1994 Los Angeles earthquake many buildings and bridges collapsed, buckled and swayed. Ten hospitals in the area were so severely damaged they were evacuated, but the University of Southern California's teaching hospital was protected by its LRBs and was unaffected. Patients did not even realise there had been an earthquake, despite the widely held belief that the seven-storey building's L-shaped design would prove unsafe. Instead, the seismic isolators it was built on allowed the ground to shake violently underneath while it remained calmly afloat. Objects did not even fall off the shelves. The ground acceleration under the hospital measured .49g and the rooftop acceleration was only .21g. That means a reduction factor of 1.8. Other buildings suffered not a reduction from the ground to the roof, but a magnification of up to three times.

When our own devastating earthquakes shook Christchurch in 2010 and 2011, there was only one major building protected by them – Christchurch Women's Hospital. It escaped unscathed, unlike so many of the city's buildings.

The shock absorbers are a great success – but there is a tragic irony to the story. In New Zealand a number of buildings are protected by the isolators – Te Papa, our national museum; the Wellington Central Post Office; and even our Parliament Buildings. However when devastating earthquakes shook Christchurch in 2010 and 2011, there was only one major building protected by them – Christchurch Women's Hospital. It escaped unscathed, unlike so many of the city's buildings. How many buildings might have been saved if Robinson's invention had been used? Even today, with the significant rebuild work going on, a number of the developers are choosing not to install the isolators because of the added cost. There is no insurance benefit or saving in New Zealand from having them installed, even with their proven effectiveness.

Robinson was awarded many honours, including the Rutherford Medal for Technology, the Michaelus Medal, an honorary DSc New Zealand, a Fellow of the Royal Society, and the Royal Society Gold Medal for Technology. In 2007 he was appointed Companion of the Queen's Service Order for services to engineering. His patents for the LRBs lapsed a few years ago but the company he founded, Robinson Seismic, continues to develop his ideas – the latest versions being isolators for use under homes, the RoGlider.

12. Keeping Things In and Out

FENCES, LOCKS AND LIDS

Each of the inventions listed here is created to protect or exclude. Maybe it is our island mentality that encourages us to make things that will keep other things away, but there are a lot of Kiwi inventions that fit into this category. Early in our European history, pioneers had to deal with the problem of keeping stock where it was supposed to be. A large country was being tamed and the geography demanded a hell of a lot of fencing.

The wire and batten fence, in various incarnations, served our fencing needs for the first century or so of settlement. There was the ubiquitous 'Taranaki gate', a concoction of wire and wood that seems to have no confirmed lineage, and indeed appears to have grown up simultaneously in many places around the world, being variously dubbed 'the Australian gate', 'the Waikato gate', even 'the Channel Islands gate'. It obviously did the job. So what prompted any new invention, if not the necessity for change? For New Zealanders, the idea that there might be another possible (and possibly better) way of doing something, even without any real need for it, is enough. For us, necessity is the mother of invention, but certainly possibility is the curious father, fossicking around in the middle of the night where he probably shouldn't be. Now, invention doesn't need three parents, so maybe we should just say that, in this part of the world, No. 8 Wire must be the milkman of invention.

The Electric Fence

ITCHY HORSES TURN WIRE INTO POWER – AND MONEY

100

To date, Gallagher holds hundreds of patents and leads the world in electric fence technology.

Right: Gallagher's modern portable electric fence.

One of New Zealand's greatest contributions to the world of agriculture came with the combination of two quite different technologies – fencing and electricity. Imagine a fence that did not rely on sheer strength for its effectiveness. This allows the fence to be light, and therefore easily portable. And a portable fence, easily moved around the farm, opens the way to revolutionise agriculture with brand-new grazing practices.

The electric fence is an integral part of New Zealand life. You've probably experienced being on the cow's end of it, climbing over it without touching it, saying 'Do you think it's on?', testing it with a blade of grass, trying to touch it between pulses, hearing the metronomic ticking as it interferes with the radio reception, or watching Wal in the *Footrot Flats* cartoon as he straddles the fence, his gummies sinking into the mud. It's fitting then that as well as being part of our cultural DNA, the electric fence was born here.

In the 1930s Alfred William (Bill) Gallagher (b. 1911, d. 1990) developed a plan to keep his horse Joe from scratching itself on the family car (an Essex, incidentally). Gallagher cunningly connected the car to an electrical supply. When the horse rocked the car, looking for a good scratch, a triggering device sent a current through the car, and through Joe. It worked. Like us, animals will go to great lengths to avoid electric shocks. Gallagher did not go on to patent his electric car protector (we can only speculate why not – perhaps because it was total

It was that Bill developed the idea from that seed to a successful electric fence product which was later to return to America, and elsewhere, as the best electric fence technology in the world.

overkill), but he did begin experimenting with electrified fences – and not just around cars, but around paddocks too.

The idea wasn't originally Gallagher's. 'I read in America where they were using electrified wire to hold stock.' But it wasn't the original idea that mattered in this case. It was that Bill developed the idea from that seed to a successful electric fence product which was later to return to America, and elsewhere, as the best electric fence technology in the world.

To begin with, Bill Gallagher linked about a mile of fence on his Waikato farm to the mains power supply. It worked fine, but laws at the time prevented the mains power supply from being used in this way. Gallagher developed a battery-powered version and by 1938 he was selling them. 'The neighbours wanted these things, so I made them some; I'd make half a dozen and go out on the road, sell them, leave them on a month's trial, and people would mostly pay for them.'

After World War II, Gallagher began full production of his electric fences. With a small staff operating from an engineering workshop in Hamilton, he made not only the battery electric fence units, but other farm equipment like fertiliser spreaders, cow bails, hay barns, cattle stops and tractors made from old cars.

In 1961 New Zealand law was changed to allow electric fences to be connected to the mains electricity supply. At that time, Ruakura Agricultural Research Centre scientist Doug Phillips took this opportunity to begin experimenting in technology for mains-powered electric fences. At first he thought he should electrify the ground! Not surprisingly, that turned out to be unworkable, so he developed a low-impedance electric fence with a very high surge of power that came on for the briefest fraction of a second each pulse. The problems battery-powered fences had with shorting on vegetation were solved, because the high power just burned any vegetation off the wires. The new technology extended the possible length of electric fences from just a few hundred metres to tens of kilometres, at the same time reducing costs. Phillips' new solution was a true worldwide first. It made electric fences encompassing whole farms feasible – and it is the basis for all modern electric fences. The Gallagher team (along with other New Zealand fence manufacturers) were quick to recognise the brilliance of the idea and adopt it.

The company began exporting in the late 1960s, and saw immediate success first in Australia, then in the UK, France and the US. During the mid 1970s the company doubled its size and production every year. Innovation hasn't stopped at Gallagher Group with success; not even with Bill's death in 1990. They developed the first set of electronic scales that can weigh cows as they walk over it, an automated sheep-weighing and drafting system, and a portable, all-in-one electric fence that pulls out like an extendable clothesline. In 1988, the Gallagher Group diversified into the security market, using their expertise controlling animals to do the same with 'two-legged livestock' – first with electrified security fencing, then with other security devices like ID tags. Then their ID tag technology completed the round trip as Gallagher began making electronic ID for stock control.

To date, Gallagher hold hundreds of patents and lead the world in electric fence technology. Not only sheep and cattle, but antelope, kangaroos and zebras are among the animals contained by New Zealand electric fences around the world. Gallagher have also sold thousands of kilometres of electric fences to Malaysia to keep elephants out – or is that to keep elephants in? Basically to keep elephants.

Now under the command of Bill Gallagher's son Sir William Gallagher, the company continues to lead the way in New Zealand manufacture and export, exceeding $200 million in revenue, sinking 7 per cent of that back into R&D, and winning awards for manufacture, design, and exporting excellence.

Not only did the electric fence prove to be a fine New Zealand export, it also encouraged innovative farming in this country. During the 1960s and '70s, the electric fence began to change farming practice, reducing the cost of growing animals, improving pasture, reducing weeds and saving millions in fertiliser. New Zealand became the best and most efficient farming nation in the world. And if you want to argue with that fact, you can argue with Bill Gallagher and his electrocuting car!

ZAMMR Handle

INVENTOR SAYS, 'IF I'D HAVE MAPPED OUT A PLAN, IT COULDN'T HAVE GONE MUCH BETTER.'

Erecting a temporary electric fence is fiddly. Sometimes you need a live connection, sometimes an insulated one, and the connecting handles would break 'as soon as a cow went near them'. Grant Pearce of Galatea in the Bay of Plenty turned frustration to opportunity with his ZAMMR Handle.

Pearce's idea was for a sturdy multi-purpose handle that could either connect the fences while conducting electricity, or connect and insulate, and could also be used as a gate-break. He made a model from clay and wire and enlisted a Christchurch company to turn that into a flash-looking prototype. 'A lot of the farmers I know like iPhones, so I thought it should be a little bit funky.'

Pearce's advice to inventors is, 'Do not fall in love with your invention.' He says the day he was waiting to hear the per-unit cost from the manufacturer he was genuinely scared. He had a cost figure written down, above which he knew he'd have to walk away and a 'Yahoo!' figure, below which he could make money on the handles. The quote came in at half the 'Yahoo!' figure. 'I ordered 10,000 and took them to Fieldays.' To this day Pearce is not sure if he would have been strong enough to walk away if the numbers hadn't stacked up.

Pearce turned up at Fieldays $55,000 in the hole. 'If that first farmer had gone, "Nah, that's not worth it," I would have been screwed.' But Pearce had tested the concept with a few close friends and felt sure it would be well received, and he was right. He sold 3000 the first day and securing a deal with big agritech manufacturer Tru-Test Group. They not only paid him for his invention but now give him a royalty for each one sold. Pearce says bringing an idea to market is 'not for the faint-hearted' but – though he won't be made a millionaire from the ZAMMR Handle – he's extremely happy with how it's gone.

The Eze Pull

A FARMING STAPLE

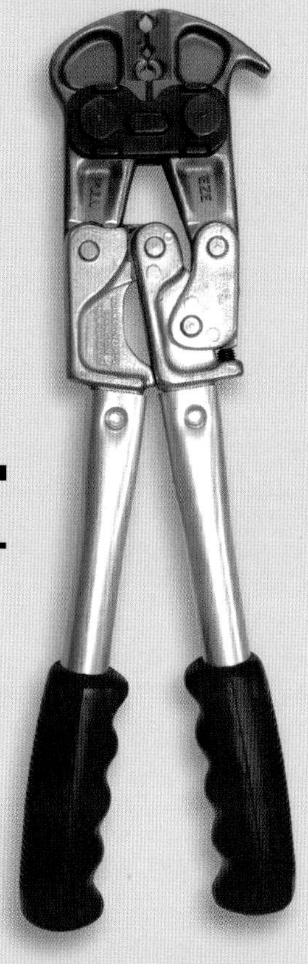

In New Zealand there are millions of kilometres of fences. On every kilometre of fence, there are hundreds of batons and posts. On every baton there are at least five staples. That means at least five hundred million staples. It also means there is a problem that needs solving – how to pull those staples out. In a farm shed in the Waikato in 1975, an old post office wire crimper was modified to make the first innovation in staple-pulling technology the world had seen in a century. Chris Johnstone was a farmer and a bagpipe player, and also had a kind of head start – he was the great-nephew of Ernest Rutherford. He put his genetic predisposition for splitting to good work and created the Eze-Pull. It is so effective that you can remove even the most reluctant staple. 'If you can see it, you can pretty much get it out.'

The Eze Pull has won every award that can be won in the National Fieldays award system, and Johnstone himself sometimes can't believe that it has all come together so well. His company has been marketing the Eze Pull internationally for over 20 years and have sold tens of thousands of the tools to the UK, Ireland, France and Scandinavia. But with all that success, Johnstone remains modest: 'I'm just a farmer who is handy with his hands. I see that as an art form.'

The Staplelok Fencing System

DO FENCE ME IN

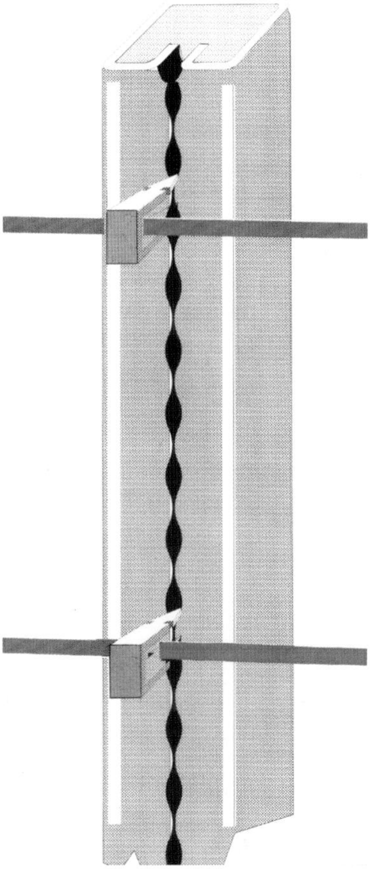

Right: The Staplelock system in use.

The story of the Staplelok fencing system starts with a topic that is often foremost in New Zealanders' minds: 'What is wrong with Australia?' In 1990, Kevin Joyce was travelling through the Australian outback and eventually felt a need to go to the loo. Stepping out of the car, he looked up and down the road and saw the warratah fence line stretching out to the horizon both ways, and thought to himself, 'There must be a better way . . .' – not to go to the loo in the outback, but to make fences. Kevin came home and with help from technicians at the University of Auckland developed a system that would have all of the advantages of the warratah, with none of the disadvantages. Staplelok fenceposts are a high-tensile, galvanised steel section, with a scalloped groove down one side of the post. The staples that hold the wires to the fence are banged into the groove with an ordinary hammer and lock into place. They will not come out until removed with the Staplelok staple remover. In fact it would take between 300 and 400 kg of force to pull the staple out without the staple remover. The posts are extremely strong and very light (2.2kg) compared to fenceposts – a farmer can carry up to 20 at one time (remember even Colin Meads could only be expected to carry one or two wooden fenceposts at a time). MAF did a cost analysis comparing normal fencing to Staplelok fencing and figured out that the savings in labour would more than cover the extra cost of the Staplelok system. The Staplelok system is a world first and Kevin Joyce and his company secured international patents (which were 'bloody expensive') on the staple locking system. Staplelok have sold the licence to manufacture the product to companies in the UK, Czech Republic and even Sweden, where they make fences to stop elk getting onto the roads.

The posts are extremely strong and very light (2.2kg) compared to fenceposts – a farmer can carry up to 20 at one time.

Below: A detail of Reid's invention.

Wire Strainers

'AN ORDINARY LAD COULD WORK IT WITHOUT MUCH EXERTION.'

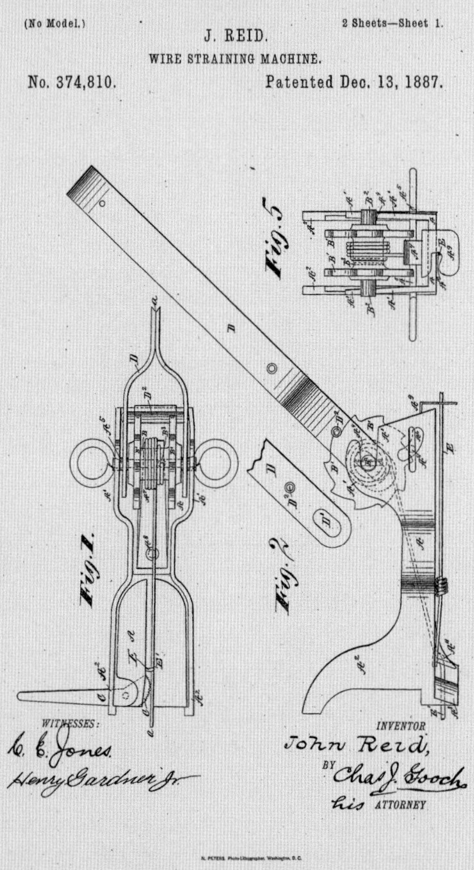

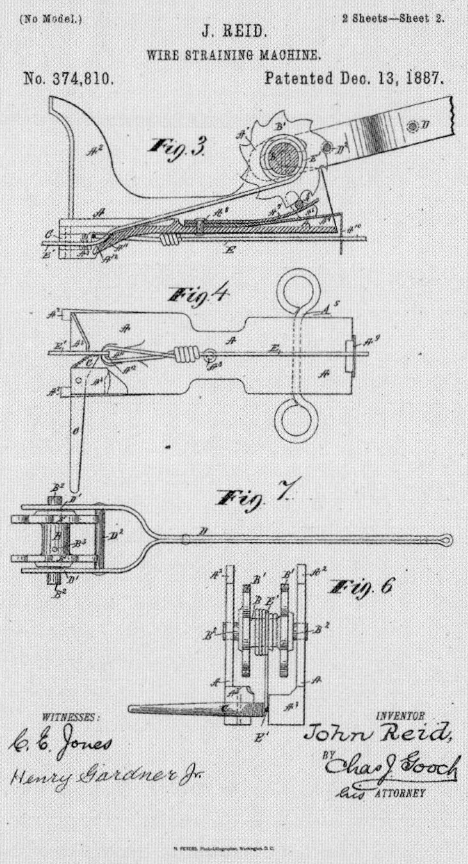

Stringing up fences around the farm then fixing them when they break is a constant occupation for farmers. One of the battles is pulling the wire taut enough so that when your cows rub their hairy behinds on it, it doesn't give too much or break. In early New Zealand history, as settlers steadily converted the land to pasture, they needed a way to carve it up and wire fencing was the preferred method. By 1878, 77 per cent of New Zealand's fenced area was enclosed with wire; and even higher in the South Island. Two engineer/farmer/inventors of that time came up with a number of ideas to make the lives of the fencing farmer easier.

John Stuart Reid (b. 1857, d. 1894) of Dunedin came up with an ingenious invention to not only fix the broken fences, but also to solve another issue. No. 8 Wire, so favoured of farmers at the time, had a couple of challenges. While it was thin enough to be

77

By 1878, 77 per cent of New Zealand's fenced area was enclosed with wire. Two engineer/farmer/inventors of that time came up with a number of inventions and innovations to make the lives of the fencing farmer easier.

Left: The strainer in action, straining.

Reid patented it in New Zealand and overseas, and achieved good commercial success with his strainer, although he also had to fend off another Otago local in the process.

pliable, it was affected by hot weather, which caused it to stretch and become loose. Farmers then had to keep tightening and loosening the wire when the weather changed. Painful.

Around 1885, Reid developed a device that could go in as part of the fence, which allowed the farmer to adjust the tension of the wire easily, and do it without the hard work of having to pull and yank the wire by hand. The device looked like a small handle connected to a wire ratchet wheel. Pulling the handle turned the wheel, which squeezed the wire through and gripped it so it couldn't slip back. It was a major labour-saving device, and soon Reid's Titan Wire Stretcher was a great success, with the papers of the day cooing over it, saying, 'The knowledge of how to use it could be acquired in a few minutes and an ordinary lad could work it without too much exertion.' Reid patented it in New Zealand and overseas, and achieved good commercial success with his strainer, although he also had to fend off another Otago local in the process.

Ernest Hayes (b. 1851, d. 1933) became well known as the founder of Hayes Engineering in Central Otago. He was born in the UK, but settled in New Zealand in 1882 and brought his engineering skills with him. He created a number of innovations – a method to poison the rabbits scourging the local farms was his first – and helped out local millers and goldminers. He created cattle stops and farm windmills, and his patented 'gold dredge crevice bucket' claimed to 'pare off a Māori bottom, whereas the present buckets in use often when let down stop the engine, owing to the large area of bucket surface coming into contact with the bottom.' We will let you try to work out what that means. He also developed a wire strainer method, the Hayes Permanent Wire Strainer, which many sources credit as the first of its kind. It appears, though, that Reid's was the first. As a further blow to his credibility, Hayes apparently also preferred the No. 11 gauge wire instead of our old friend No. 8 (see page 268), as he claimed it was lighter and more cost-effective. However, he can be credited with the creation of the Monkey Brand wire strainer, which deserves a mention for the name alone. Hayes and his sons went on to invent a number of variations to the wire strainer to improve it and make it the ubiquitous star on many New Zealand fences.

Wire strainers based on Reid's original design, and with enhancements from Hayes, are still manufactured to this day.

Hamburger Holder

NO MORE EMBARRASSING STAINS

In 2002 New Plymouth-born Roland Matthews came up with a novel invention for holding his hamburger. Driving to Auckland one day, he got tired of having a McDonald's Mega Feast landing on his lap – and the idea for his folded paperboard packaging was born. After searching international patent records, he realised that in the more than 100 patents for hamburger packaging, none solved the issue of mayonnaise dripping on your pants. Matthews underwent a rigorous burger-holder design and testing procedure, and came up with a simple folded-cardboard design that cups the burger as you eat it. When he was happy with the design, he showed his creation – the imaginatively named Hamburger Holder – to New Zealand burger chain Burger Fuel. They loved the idea, licensed it from him, and put it on all their counters as the Doofer. The product is now used throughout New Zealand and in Australia, and now Matthews is taking it global. The world is his burger.

After searching international patent records, he realised that in the more than 100 patents for hamburger packaging, none solved the issue of mayonnaise dripping on your pants.

Below: Another satisfied customer of Roland Matthews' Hamburger Holder.

Sector Navigation Lights

PROVIDING SAFE PASSAGE FOR BOATS

Below: The sector lights mark the safe passage into tricky harbour entrances.

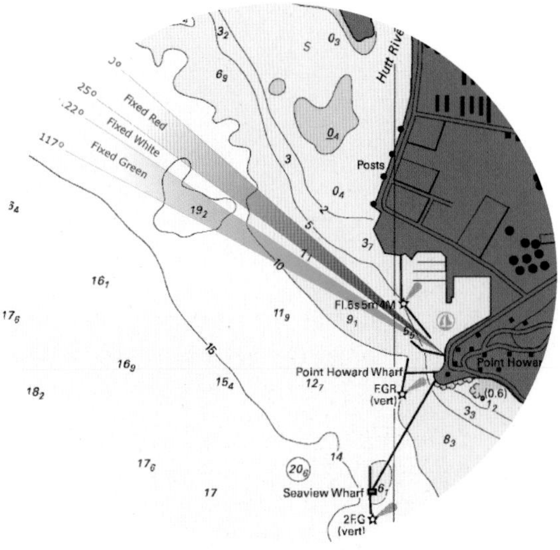

It was a simple problem, but a deadly one. The entrance to the Pauatahanui inlet of Porirua Harbour is full of sand-bars, rocks and a reef that will sink a boat quick as look at it. Further, the traditional way of marking out the safe entrance to the harbour was unsuitable.

In the past, boats were guided into harbours using a system of two lights – one above and quite a bit behind the other. When the captain sailed a course so that he could always see the two lights atop one another, he would know he was on course.

But in Porirua Harbour, as in others throughout the world, there is a large cliff near the entrance at Goat Point, exactly where the rear light needed to go. For many years Porirua Harbour stood unmarked and treacherous, especially at night.

It was this problem that the committee of the local boating club tried to solve when they met one evening in 1961. One of the members, Bob Barnes, took the problem to work with him the next day and asked his workmates. If he had been working at Pita Pit we probably wouldn't be talking about it now but, luckily, he worked at the Department of Scientific and Industrial Research (DSIR).

Barnes' colleague Norm Rumsey, an optical physicist, immediately saw the answer. Literally by lunchtime, he had the working prototype of a solution. Rumsey's invention was to be a boon for harbours worldwide, and a successful local export.

Norm created a kind of modified slide projector, which shone three beams of light across the dark harbour. A very narrow white beam in the middle was flanked by a beam of red to one side and a beam of green to the other. When approaching the harbour now, the captain of the boat simply ensures he can only see the white light – if he sees red, he's too far left; green, too far right.

Rumsey's experience designing high-quality lenses enabled him to design lenses that could shine the narrow beams of light for kilometres without dispersing uselessly. Vega Industries licensed the 'sector navigation lights', named them PEL (after the Physics and Engineering Lab at the DSIR) and turned Rumsey's invention into a million-dollar-a-year business.

The PEL system now guards harbours and waterways as far-spread as California, Angola, Durban and Diego Garcia – and even helps large ships pass each other through the tight bits of the Panama Canal. Vega are also making guidance lights for airports using similar principles.

Back in the Porirua Harbour, Norm Rumsey's original light system kept marking out the safe channel for 30 years, before being replaced with the latest version.

Childproof Lids

KEEPS LITTLE FINGERS OUT OF YOUR STUFF

You know the ones – the caps on medicine bottles that are ostensibly childproof, have foxed the odd unsuspecting adult too. They are the plastic caps you have to push down then twist in order to open them. They've sold in their tens of millions worldwide, and they were invented in Whanganui.

Claudio Petronelli is an Italian expat who moved to Whanganui and had a family. In 1970 he had a baby son, Antony, and another on the way. Reading the newspaper one day he came upon the story of an unfortunate family in Hawke's Bay whose toddler had found the family medicine cabinet and swallowed a cocktail of drugs, with fatal effect. As a father, Claudio could empathise with the family. He set about inventing a childproof lid system – or, as the patent application put it, 'A safety closure with an outer ring mounted over the threaded cap for a bottle with an annulus extending in from the outer ring engageable in a groove on the outer surface of the cap and with a clutch engagement between the

10

They are the plastic caps you have to push down then twist in order to open them. They've sold in their tens of millions worldwide, and they were invented in Whanganui.

Left: Inventor Claudio Petronelli demonstrates his garlic mincer for a lady in a stripy dress.

They licensed the manufacture of their childproof lids to an American company, who manufactured millions of them.

annulus and the side of the groove at one limit and preferably the upper limit of axial movement between the ring and the cap.'

Let's agree to keep calling it a childproof lid.

Claudio was an engineer by birth; his father was a civil engineer in Rome, and we all know the reputation of Roman civil engineers – their roads and aqueducts were the best in the world! Claudio made a prototype cap in metal, coming up with the double-lock system that we know today. With partner Gavin Park he refined the prototype until he had a working model, which they then began manufacturing in plastic. They registered their patent, a process they describe as very costly, and then set about licensing their idea.

The patent to their design has now lapsed, but while they had it, they licensed the manufacture of their childproof lids to an American company, who manufactured millions of them. While Claudio made some money from the idea (and in fact most of his inventions), to him the greatest satisfaction is knowing that his system is protecting millions of children worldwide. When the lids were introduced in the United States in the 1970s, there was a dramatic reduction from the 82,000 kids a year who were sick or even hospitalised after eating pills they shouldn't have.

Claudio, unlike his Roman ancestors, didn't rest on his laurels. He formed Petronelli Industries, a manufacturing and design company based in Whanganui. The company has created a number of other inventions, from garden tools (the trowel fork is one of theirs), to kitchen utensils (an improved garlic mincer), water valves, an improved manifold and even the box thing that surrounds water tobies.

Claudio doesn't believe that being based in a small town, or even in a small country like New Zealand, is in any way a disadvantage to the inventor. But he does caution that inventing can be an expensive business and – from the initial idea, through prototyping, manufacture, patent applications, and legal costs to seeing an income from your idea – a long haul. Today he works in his own hairdressing salon in Whanganui, but even while he's cutting hair, he's always thinking and his mind is constantly inventing.

BFM Fitting

A SMALL BUT PERFECTLY FORMED REVOLUTION

5

Five years since its birth, the BFM Fitting is sold in 100 countries around the world.

Right: Blair 'middle name probably begins with F' MacPheat.

The BFM Fitting is a New Zealand invention currently developing a very nice niche market around the world. The inventor, Blair McPheat, had an advantage from the very beginning. From the age of nine in 1977, he had been working with his father, who manufactures and sells industrial equipment. As soon as he was old enough, he was on the road selling the product too and, he says, 'My customers were screaming and yelling that they were having so many issues with connectors, particularly hose-clip type connectors.'

Every processing plant, whether it be food, pharmaceutical or petrochemical, has connectors between the different components – the hoppers and vats, sifters and conveyers etc. These components are often in motion, so the connections need to be flexible. The incumbent connectors were often just any old piece of material attached on the outside of the pipe with hose clips. 'You'd be amazed at all the high-tech, million-dollar machines around the world with grotty old bits of horse blanket and a hose clip.' The inevitable result is chronic leakage, breakage, wasted time and dust everywhere.

After months of pondering the problem, McPheat sat up in bed at 2am in his Devonport, Auckland home and told his wife he was going to work. He'd had the idea which was to lead him to his new business.

The BFM Fitting he dreamed up is a sleeve of flexible and see-through urethane with a strong but flexible snap ring integrated into each end. To fit it you just squash the ring with your hands, pop it inside the pipe it's connecting to, then let it go and it seats itself. Taken as a system McPheat's invention revolutionises industrial connectors – giving a perfect, tough seal that will not leak and is easy to remove and replace.

The next few months were hectic. 'There was a fair bit of trial and error' as McPheat struggled to get the fitting manufactured the way he saw it in his head. A salesman at heart, he was champing at the bit to get out and sell it, but knew he couldn't until it was right. Still, within a year of his eureka moment, he was on the road.

His customers were thrilled. Most said it was the simplest thing they'd ever seen in their lives. Five years since its birth, the BFM Fitting is sold in 100 countries around the world. The company holds patents in almost every territory and their global reach grows every month.

Let's finish with McPheat's advice (apart from 'go to work at 2am'): 'Have the right product. Don't refine something that's already there. Make sure that you are putting effort and money into something that has a good point of difference. Product is the most important thing. See a need, fill a need.' Fitting.

Reusable Envelope

RIP IT, PEEL IT, RESEAL IT

50

With customers like Telecom, Vodafone, Greenpeace and the ANZ Bank, Candida make over 50 million of them a year.

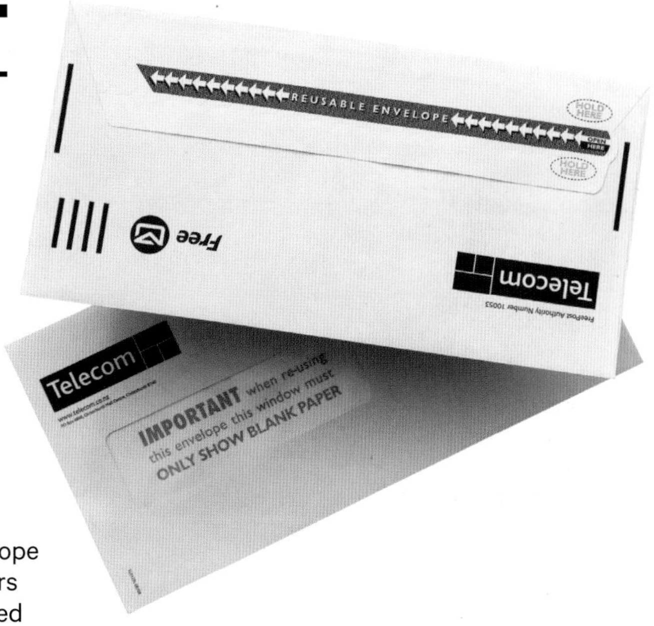

Right: The envelope please…

If you ask Stephen Smythe if the reusable envelope he invented was the first in the world, he answers truthfully, 'One never quite knows.' Smythe retired from his architecture practice at the end of 1997, and with time on his hands an idea that had been germinating in his head sprouted shoots.

Remembering how his father, affected by the Great Depression, had reused envelopes by removing the stamps and crossing out the address, then resealing them with tape or staples, Smythe decided to invent a reusable envelope.

By 1999 he had a design and a patent. It was a genuine first. There are two envelope manufacturers in New Zealand. Croxley didn't want to know but Candida were interested and luckily they could make Smythe's design on their existing machines. The first crowd to give the envelopes a go was the North Shore City Council, who sent bills out using them. Smythe well remembers the day in 2000 when his first one arrived in the mail – it was a disaster. Someone at the council had shoved too much into the envelope and it was bursting open. 'Standing there at the letter box my heart sank and I thought, "That's it, they've killed it."' But the good customer feedback outweighed the small problem and the Echo envelope was away.

Echo envelopes are now familiar household objects. With customers like Telecom, Vodafone, Greenpeace and the ANZ Bank, Candida make over 50 million of them a year. Although Smythe has patented it overseas too, he's not wealthy yet. 'As most inventors will tell you, you're so keen to see it up and out the door, that you just do the deal that gets them going. It's not a big earner.' But he isn't afraid to say he's terribly proud of making it happen, and he rates New Zealand as a fantastic place to get a new invention to scale. 'If I'd done this in Britain or America, it would have died a horrible death. Getting stuff up and running in New Zealand is a cakewalk compared to other countries. In New Zealand we are interested in making new stuff and we like the idea of helping other people. And here you know everyone, and you can talk to the person making the decisions. When we went to New Zealand Post they ran a trial and they said, "This is a dog – it doesn't work." I said, "You've got it upside down." Overseas it would have ended there because I wouldn't have been in the room.'

Vacuum Mooring System

MONTGOMERY'S SHIPPING SUCKERS

Firstly, if you know what a hawser is then you probably know all about this invention. If, though, like most people you've never contemplated the mooring action of a large ship and didn't realise a hawser is the proper name for those really big ropes they use to tie ships up, then read on. You will learn about a Kiwi invention that saves time, money and lives – and is now successful the world over in its field.

Peter Montgomery – no, not that Pete Montgomery, a different one – is a serial entrepreneur. He's had eight start-up companies, and with a track record of three successes, three okays and only one huge failure, he reckons he's done all right. He always starts with a vision of the future – although he does caution that sometimes the visions are hallucinations and it's important to be able to tell the difference.

600

He amassed around 600 rejections before a lucky turn of fate saw him with success right in his own backyard.

Left: Ironically, after trying all around the world, Montgomery's first customer was right here in New Zealand.
Right: Montgomery's mooring system sticks ships to the dock.

A safer way to moor boats was one of his visions, sparked after seeing the death of a ship worker when a mooring cable snapped and the end of it hit him in the chest. Montgomery's system replaces the hawsers with a vacuum system – literally sticking the ship to the dock with suction. It's also faster, cheaper and uses less labour.

However, the shipping industry is pretty conservative and in 1996 when he tried to get ship owners and shipyards interested in his design, they thought it was pretty far-fetched. He amassed around 600 rejections before a lucky turn of fate saw him with success right in his own backyard. Almost in desperation, he faxed a copy of his plans off to Tranz Rail. He didn't really expect anything to come of it, but they rang within the hour and he met with them the next day. Coincidentally, they had been looking for a new mooring system for their new Interislander ferry, the *Awatere*, and had even made allowances in their plans for the as yet unknown system. From this serendipitous situation Montgomery established his first working system, which stayed in service for 10 years before being replaced by one of his newer designs. It was a lucky intervention – Montgomery was 'in the last 10 yards before giving up the race' and the opportunity for a working system so close to home turned near-failure into huge success.

He ended up floating his company on the New Zealand Stock Exchange, then working closely with international engineering firm Cavotec, who helped him create the second-generation system, a shore-based design. Cavotec eventually bought out the New Zealand firm in 2006 and retains the worldwide rights to the MoorMaster system – a system now in use on docks around the world. The MoorMaster sucks a ship to the dock with over 40 tonnes of vacuum pressure. It can secure and release a ship in 10 seconds, with far less manpower and much more safety than the old system of ropes (sorry, hawsers).

Cavotec eventually bought out the New Zealand firm in 2006 and retains the worldwide rights to the MoorMaster system – a system now in use on docks around the world.

Montgomery stuck around for a few years in Cavotec, then left the venture to get back into the start-up game. He's currently working on a product to improve safety in the general workplace, with Clever Medkits, a system which is like a first-aid box with a brain – it monitors its own inventory, helps users select the right treatment, and allows remote communication with medical professionals. He's about to start trials with a very large multinational customer. Montgomery hopes that Clever Medkits will do for the general workplace what MoorMaster has done for the shipyard – create a safe and fast system, with global business appeal. Hawsers begone.

Securichain

A CHAIN FOR SECURING DOORS

14

Murray and his team set themselves the task of making a door-chain that could withstand 12 shoulder charges by a 14 stone (88kg) man

Right: The bane of the burly burglar, the Securichain.

In 1982 Murray Baber was an inventor looking to make a money-spinning business from something that would sell like hotcakes. As a locksmith familiar with door-related products, it was natural that he would come up with an idea to improve the familiar security door-chain. But what could possibly be wrong with the common door-chain? At first glance, nothing. But that is the inventor's genius – to create a small improvement then print cash. Baber identified three problems with the traditional door-chain. It is ugly and isn't suited to the décor of a lot of modern apartments or hotel rooms; it is fiddly for elderly or disabled fingers to operate; and it is weak – if you kick the door the screws holding the chain to the architrave just tear straight out.

So Baber designed a better door-chain. The Securichain is fitted *inside* the door and the jamb, so that when it is closed the only part that is visible is a little box unit that sits on the edge of the jamb with a switch allowing you to easily engage or disengage the chain. The chain is like a little bicycle chain, making it very strong. Baber and his team set themselves the task of making a door-chain that could withstand 12 shoulder charges by a 14 stone (88kg) man, or presumably one shoulder charge from a dozen 14-stone men – thus preventing most, but not all, of the All Blacks from breaking into your house. They enlisted the help of DSIR to test the unit, and kept refining the prototype until it was strong enough.

In order to make the unit stylish, Baber enlisted the aid of Peter Haythornthwaite, an industrial designer. Baber rates this as one of the best decisions he ever made. The product won New Zealand's top design award in 1988, as well as an inventors award. Baber's company joined forces with an established company, Interlok, to market the Securichain, supplying the product to hotels all over the world. Then in 2001 one of the world's largest lock companies, ASSA ABLOY, bought Interlok including Securichain. The same year, Baber's patent expired and companies in Asia began to copy the product at a cheaper price.

Baber doesn't mind. He's moved on and is still developing innovative security products. His success has come not just from a brilliant idea, but from good business sense, great design and strong financial backing. He admits he's a serial inventor: 'Looking for the next best mousetrap keeps me going.'

Kaynemaile

LIGHTWEIGHT PROTECTION FOR BUILDINGS

Kayne Horsham was art director for creatures, armour and weapons for the *Lord of the Rings* film trilogy when he realised the fake chainmail that his team was creating could be made a different way. He worked out how to create a 2D mesh of solid components that wouldn't suffer from the same breakages he was seeing in the fake armour being made. Taking the name that the actors used to use to describe this new fake chainmail, he created Kaynemaile – a unique seamless polycarbonate mesh.

Horsham quickly realised his invention had applications far beyond fake chainmail. He started a company and is now making Kaynemaile in large quantities for architectural design in building facades, screens and curtains. It's even been adapted as a type of shark net. The product looks good and functions well, and has rapidly become a key feature of some great buildings around the world –

The product looks good and functions well, and has rapidly become a key feature of some great buildings around the world.

Hard Rock Cafés through the US, Cornell University in New York, and Air New Zealand's check-in areas and lounges all use it. Horsham has done all this while remaining a proud Kiwi company: 'New Zealand companies are often advised to find a partner offshore to commercialise their offering but we've found, with persistence, you can sell direct in global markets. And the benefit is that you know who your customer is, the way they think and what they want. That drives improvements we make to the product.' It also repels hordes of orcs.

Below: Kaynemail also enhances the look of building exteriors.

Multiple-kill Pest Traps

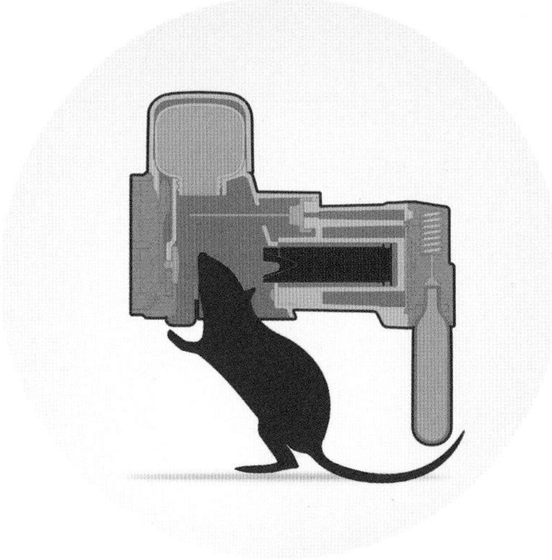

HUMANE KILLING FROM RATS TO MINK

This invention takes the prize for irony – it is at the same time the most humane invention in this book and yet the most delightfully brutal. The problem with possum and rat traps is that they can only catch one animal at a time so they need to be constantly reset. The only alternative is poison, which is controversial. Goodnature possum, rat and stoat traps, made in Wellington, are the world's first commercially available traps that kill multiple vermin without resetting and without poison.

First the rat or possum is lured by the smell of some sweet bait mixture. There it is, inside that thing attached to the trunk of this tree! Gripping the tree trunk, the possum pokes his pointy head up into a cone-shaped plastic portal with the lure at the top. He gets the barest taste of the lure before he triggers a punch, powered by a CO^2 canister, which gets him in the back of the skull. He never knows what hit him and falls out of the tree – dead before he hits the forest floor. Then the trap automatically resets.

It's not uncommon for one trap to account for five possums in a single night, and the CO^2 canister is good for 20 kills. Because possums are only a pest in New Zealand, Goodnature have made a version that nails rats and stoats as well. In a trial of two rat traps in Indonesia each trap killed seven rats in one night.

Goodnature's innovators Craig Bond, Robbie van Dam and Stu Barr met while studying industrial design at Victoria University. They started the company in 2005, scored some crucial funding from the Department of Conservation in 2007, went full time in 2008, and they've sold over 12,000 traps so far. The potential market is global. Overseas, they're working with the Swedish government on traps to kill pesky mink. They are also developing traps to help Hawaiians kill mongooses (mongeese?), and the huge and extremely destructive rat population of Asia is in their sights.

With an emphasis on great industrial design, Goodnature have built a better rat trap and they're beating a path to the world's door.

Above: Pop goes the weasel (and rat, stoat, etc.).

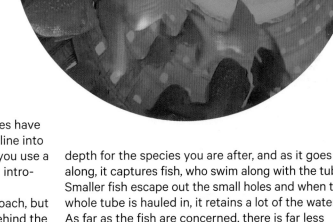

Below: Little do they know, that's a blue tube of piscine death . . .

Humane Trawling

A NEW APPROACH TO AN OLD WAY OF FISHING

For centuries, the same two basic techniques have been used to catch fish: you either throw a line into the water, with one or more hooks on it, or you use a net. A New Zealand company is working on introducing a brand-new third option.

Modern trawler fishing uses the net approach, but on a grand scale. Huge nets are dragged behind the boat, often scraping the bottom and certainly always catching things that aren't expected or desired. It's an inefficient way of catching what you *do* want, bringing up a net full of stuff you *don't* want, then dumping it on deck and sorting through the catch. Millions of undersized fish, fish of the wrong species and other marine animals are dredged up in this way, not to mention car tires, gumboots and other ocean flotsam.

Scientists from Plant & Food Research, partnering with three large New Zealand commercial operators, have come up with a way to avoid that inefficiency and reduce the impact on fish stocks caused by trawling. The precision seafood harvesting technology they have developed at a cost of over $50m is set to revolutionise the fishing industry, and with the patents in place, they hope to see global success for their ideas.

The system works by using a giant tube, rather than a net, with small holes or escape portals in it. The tube is dragged behind a boat at the desired depth for the species you are after, and as it goes along, it captures fish, who swim along with the tube. Smaller fish escape out the small holes and when the whole tube is hauled in, it retains a lot of the water. As far as the fish are concerned, there is far less trauma: it's like their swimming pool just got moved. The fish can be kept alive longer and any unwanted species can be put back in the ocean with much less handling or stress.

At the time of writing, the new system was two years into a six-year-long trial, and not ready for mainstream release. Nor were details of the specifics of the system yet available for public consumption ('commercial sensitivities'). But with $50m investment, it's clear plenty of people are betting it's going to be huge. Alistair Jerrett, part of the R&D team who created the system, said the 'aha!' moment for him was during the initial research, when they started to ask why it was necessary in traditional fishing to exhaust the fish and strain and damage them during the harvesting process. The new system they have developed allows that to be avoided for a more precision catch, a more efficient fishing system, and a more humane approach to this ancient art. Let's hope this fishy tale isn't one that gets away.

13. Myths, Legends and Icons

REFRIGERATED SHIPPING, THE BUZZY BEE AND OUR INVENTIVE NAMESAKE, NO. 8 WIRE

We New Zealanders are so proud of our innovative nature that sometimes we overstep the mark in claiming inventions as our own. All it takes is a bit of confusion early on, just a couple of people getting the wrong end of the stick, and a story can catch on, spread slowly then grow into a myth.

In this chapter we address some of those myths. These are the sort of thing you might have expected to find in this book. If we hadn't mentioned them you'd probably finish reading and say, 'The idiots didn't include trench warfare.' But while most aren't actually New Zealand inventions, they are all interesting stories involving legendary New Zealand characters. There are quite a number of other innovations and inventions that we were instructed by interested parties to investigate, which diligently we did, every last incorrect one of them. I mean, bubblewrap? Not even close to being one of ours. The cat's-eye road marker? Also not from these islands. The trench periscope, the three-colour pen, the rotary lawnmower, Velcro and the fax machine were other interesting red herrings. Someone even suggested we invented the idea of colouring herrings red.

To bolster your spirits, we also include in this chapter icons of such cultural importance that they demand to be included here, even if, though they are definitely 'New Zealand', they are possibly not really 'inventions'. Hokey pokey is just too good to leave out.

Things with the Word 'Kiwi' in the Title

WE'LL TAKE WHAT WE CAN GET

Above: A rampant kiwi adorns the 6d stamp.

This section is a cold, hard, factual look at some things with the word 'kiwi' in them. Unfortunately trying to claim any of these things as Kiwi inventions is the very definition of clutching at straws...

First of all comes the kiwi bird. It is entirely indigenous to New Zealand. Nobody else can claim it, but while we should be proud of it, it is not an invention of ours at all. It evolved by itself; in fact, it was doing much better before we came along. We will note, however, that it lays the largest egg relative to its body size of any bird in the world. Already world leading.

We will note, however, that it lays the largest egg relative to its body size of any bird in the world. Already world leading.

Next comes Kiwi boot polish. You're not going to like this. The company that made Kiwi polish started making it in 1906. It has become an international success story, expanding to Britain and the US as well as France, Canada, South Africa, Spain and Pakistan. The polish made 'Kiwi' a commonplace word all over the world. The company didn't invent shoe polish, they just made a new kind, but that's not the bad bit. Here's the bit you won't like: the company is Australian. In 1901 William Ramsay visited New Zealand from Melbourne. He married a woman from Oamaru called Annie Meek. Later, back in Melbourne, he started making the soon-to-be-famous polish and his wife suggested the name 'Kiwi'. Kiwi polish is Aussie.

Ramsay's wife didn't suggest calling the polish 'Kiwi' after the people of New Zealand, because we weren't known as Kiwis until a little later. Indeed, the first time the kiwi is thought to have been used as an emblem was in 1887, when the University of Auckland used three kiwis on its seal. In 1905, the *New Zealand Herald* printed one of the first known cartoons where a kiwi was used to represent New Zealand, depicting a huge kiwi eating the Welsh rugby team. The existence of the Australian boot polish may in fact have added impetus to the naming of our people. During World War I, the polish was widely used by

Allied troops, and Kiwi became a well-known word. By the end of the war, it was common to call New Zealanders and things New Zealand, Kiwi. But that isn't a New Zealand invention either, it's just a name.

The kiwifruit was first grown in the Yangtze Valley in China. Its name in Chinese (which it would be fair to say is its real name) translates to mean 'monkey peach' – maybe because while it's yummy like a peach, it is also hairy like a monkey. Jon has the same nickname. Monkey peach seeds from China were brought back to Wanganui in 1904 by the headmistress of Wanganui Girls' College, Isabel Fraser, who passed them on to local farmer Alexander Allison, who had an interest in unusual plants. By the 1940s and 50s, many New Zealand homes had a vine in their backyard. Because the tree was from China, and despite the fact that the fruit looks absolutely nothing like a gooseberry, the fruit was called Chinese gooseberries by New Zealanders.

The fruit would have stayed named that if it hadn't been for the fact that during the late 1950s Turners and Growers began exporting the fruit to the United States. The Americans had a problem with the name Chinese gooseberry, as it seemed to describe another fruit all together (and quite a yucky one at that). The Turner brothers put their thinking caps on and came up with 'melonettes'. Nice, yes, but anything to do with melons attracts a 35 per cent import duty in America so the name

Figuring that the word 'kiwi' was the only word that foreigners would associate uniquely with New Zealand, they renamed the fruit 'kiwifruit'.

was dropped. Figuring that the word 'kiwi' was the only word that foreigners would associate uniquely with New Zealand, they renamed the fruit 'kiwifruit'.

While New Zealanders didn't invent the kiwifruit, we did get pretty gung-ho about it. (The term gung-ho or 'work together' was coined and promoted by Rewi Alley, a remarkable New Zealander and friend of China. He remains the only foreigner ever to be given a state funeral in China.) We perfected kiwifruit by refining the plant to produce perfect fruit for human consumption. The kiwifruit has significant proven health benefits, including an enzyme which makes you feel less bloated, very high vitamin C and rapid protein absorption power. It remains a significant horticultural export, with innovation meaning new breeds and varietals have emerged (golden kiwifruit anyone?). It might be larger if it weren't for the fact that we also exported the plants so that other countries like Chile could benefit from the world kiwifruit boom. Some people see that as a big mistake, but let's look at it as a form of foreign aid.

Since we could not patent the kiwifruit, or even trademark the name 'kiwifruit', the centralised kiwifruit marketing agency has come up with an appellation for New Zealand kiwifruit that no one else is allowed to use. Just like only sparkling wine from the French province of Champagne is allowed to be called champagne, only kiwifruit from New Zealand can be called Zespri. Sadly today the kiwifruit industry is battling a virulent disease, *Pseudomonas syringae actinidiae* or PSA, which somewhat ironically, appears to also have originated from China.

Speaking of Golden Kiwi – you'll be pleased to know that the lottery of that name, as anyone old enough to have lived through the period 1961–89 will remember, was a New Zealand creation. It replaced the mysteriously named Art Unions of the previous era, and was itself surpassed with the creation of the Instant Kiwi, a scratch card lottery still popular today.

The Buzzy Bee

PULL-ALONG FUN

1

So good in fact, was the Buzzy Bee, that the future King of England, Prince William, was given one on his first trip to New Zealand, an occasion he no doubt remembers fondly, even though he was one at the time.

Right: *A Buzzy Bee for Siulolovao*, print by Robin White (1977).

In the late 1930s, Auckland toymaker and wood-craftsman Maurice Schlesinger, in a small workshop in Saint Benedicts Street in Newton, Auckland, had a blinding flash of inspiration and almost single-handedly created the basis of New Zealand's cultural and popular identity. Schlesinger designed and created the Buzzy Bee toy and thus surely became a candidate for New Zealand's most important but least known historical figure.

Forward-thinking Schlesinger used lead-free paint and a wood lathe to create the products, and sold them around Auckland. The wooden bee toy was pulled along by string and the rolling motion of its wheels made a clacking sound and spun the yellow wings. A fascinating and ground-breaking invention it might not have been, but certainly a successful product both here and overseas. More than a great product: through fond association with our childhoods, the bee has become a Kiwi icon and epitomises what we now call Kiwiana – appearing on stamps, books, paintings, posters, songs, TV and even the keel of America's Cup boats.

Unfortunately Schlesinger became very sick and so he passed on the toy design to travelling salesman Hec and his brother and wood-turner John Ramsey. In 1948 these two continued production of the toy bee, which has remained basically the same for all of its 60-plus years of production. The plastic wings

Unfortunately Schlesinger became very sick and so he passed on the toy design to travelling salesman Hec and his brother and wood turner John Ramsey. In 1948 these two continued production of the toy bee, which has remained basically the same for all of its 60-plus years of production.

were, until the late 1960s, made of the same kind of fibreboard used to make old suitcases, and the clacker mechanism was updated in 1993 to suit modern safety standards, but apart from that it's the same good old bee. It's interesting to point out that the wings were always made of plastic or fibreboard. They have never, ever, ever been made of wood. If you fondly reminisce about the old days with the image of a wooden-winged Buzzy Bee in it, then you are just plainly wrong.

So good, in fact, was the Buzzy Bee, that the future King of England, Prince William, was given one on his first trip to New Zealand, an occasion he no doubt remembers fondly, even though he was one at the time.

There has always been only one real Buzzy Bee. The numerous patents and trademarks that protect Buzzy Bee have been bought and sold a number of times since the days of Schlesinger, along with the rights to several other of his toys such as the 'Mary Lou Doll' (from 1941) and the inventively named 'Richard Rabbit'. The tradition of Buzzy Bee remains strong.

Several million bees have been made, and sales are still very strong in New Zealand, but Buzzy Bee has never been a huge export. A big part of the Buzzy Bee business in the early twenty-first century is merchandising. The Buzzy Bee company, located in sunny Warkworth, has issued over 100 licences to other firms to manufacture and sell Buzzy Bee pyjamas, lunchboxes, stamps, schoolbags, jewellery, wrapping paper, paperweights, even women's underwear.

So, a New Zealand invention, and a New Zealand success story? Maybe. Unfortunately, in the last year of last century a dark shadow of doubt was cast on the Kiwi origins of the Buzzy Bee...

A fellow named Vernon Davenport spoke up in the media about having worked in the Ramsey factory at the time of the bee's creation. Vernon Davenport (if that is his real name) says a guy called George Steel (a likely-sounding name) brought a toy bee from America and showed it to Hec Ramsey. In another version of the story, the bee importer was Ramsey's own sister (a girl called George?). Either way, the American 'bee' (which was even supposedly called 'Buzzy Bee') according to both stories 'was flat, about one and a half inches thick, timber, with the centre gouged out where the noise contraption went. It had coloured paper stuck to its body instead of paint... Hec looked at it and decided to make his own version. Within three months we were making batches of 5,000 to 10,000.'

Indeed, the US toy manufacturers Fisher-Price did create a shallow imitation of our Buzzy Bee in the 1950s that matches Davenport's description. Fisher-Price must have realised the inadequacy of their folly because they made a few versions over the years but stopped in the mid-1980s. Schlesinger's Kiwi bee design does appear to pre-date the American version, and anyway according to our patent experts, the rounded, painted bee that we know would be a sufficient departure from the alleged American bee to warrant a new patent, and to be in fact a different toy. Also, if the American bee had been any good, it surely would now be a worldwide success story, and not the historical cul-de-sac that it is.

As New Zealanders certainly we must thank the Americans politely for their 'bee toy' but ask them, for both of our sakes, to stick to electricity and space travel and leave the important inventions to us. If the Americans want our bee, I think I speak for everyone when I say they'll have to come over here and take it off us. And anyone caught helping them (you know who you are, Vernon Davenport) will be tried for treason and hanged.

Refrigerated Shipping

OPENING NEW ZEALAND UP TO THE WORLD'S MARKETS

In the 1880s, whether we knew it or not, New Zealand was in an economic fix. Writes historian Michael King in *The Penguin History of New Zealand*, 'It is difficult to see how New Zealand could have survived as a viable country solely on wool and grain and extraction commodities for its national income.'

Our commodities started off very promisingly, but the gold rushes which had kick-started our economy with an influx of money, immigration, investment and infrastructure were over. The kauri forests, which had been logged unsustainably, were almost gone. And, as if the kauri hadn't given enough, our other great commodity was kauri gum. The international hunger for kauri gum, used to make varnish and lino, would dry up almost completely by the early 1900s. At one time 70 per cent of the world's varnish contained New Zealand kauri gum, and much

Finally on 15 February 1882 the first shipment of what was to become, and remain, our largest export industry sailed from Port Chalmers on the *Dunedin*.

of Auckland's wealth came from the export of millions and millions of little golden lumps of it. But gum became harder to find and when synthetic alternatives were invented the industry was doomed.

The obvious answer was agriculture, and there was no doubt New Zealand was a pastoral powerhouse – in 1881 there were 13 million sheep in a country of fewer than a half a million people. The trouble was that the only viable agricultural exports were grain

0.5

By 1933 half of Britain's imports of lamb, mutton, butter and cheese came from New Zealand. The new industry meant not only economic prosperity for New Zealand that lasted until the 1970s, but massive social change around the country.

Left: Frozen export carcases outside the British New Zealand Meat Company in Christchurch, circa 1910.
Far left: An oil painting of the *Dunedin* by Frederick Tudgay records the first shipment of frozen meat to Britain in 1882.

and wool, and even the price of wool was slumping. Three months by sea is a long way for a country to be distanced from its largest trading partners; butter and meat would be well past their use-by dates before the ships even left harbour. Excess sheep were often herded off cliffs into the sea to get rid of them because the local market for meat was too small. For all the same reasons, there was hardly any point in growing cows at all.

All the while on the other side of the world the Industrial Revolution had led to a rapidly increasing demand for meat in England.

To the rescue came two determined New Zealanders. William Soltau Davidson (b. 1846, d. 1924) was the general manager, and Thomas Brydone (b. 1837, d. 1904) was the New Zealand manager of the New Zealand and Australia Land Company – a major landowner of large estates around the country. They began to exhaustively research and plan.

The first successful shipments of frozen meat had been made just a few years earlier. In 1877 two steamers carried frozen mutton from Argentina to France, and in 1879 the *Strathleven* carried 40 tons of frozen beef and mutton from Sydney to the UK.

New Zealand was a longer journey, but the plan was the same. In 1881 Davidson and Brydone had a ship, the *Dunedin*, refitted as a floating fridge. The Scottish Bell-Coleman steam-powered refrigeration plant on board was state of the art – the same sort of unit as used on the *Strathleven* – but the venture was by no means certain. Despite promises of fresh meat every night – previously unheard of on a three-month voyage – many passengers refused to travel because they were afraid the plant would set fire to the ship's sails. The first shipment loaded thawed out when the refrigeration plant failed: it had to be offloaded and quickly sold. But finally on 15 February 1882 the first shipment of what was to become, and remain, our largest export industry sailed from Port Chalmers on the *Dunedin*. After 98 days at sea, 4460 sheep and 449 lambs reached Smithfield Market in London safely, and every animal was in good condition (still dead, but the meat was perfectly edible).

The floodgates had been opened. By 1933 half of Britain's imports of lamb, mutton, butter and cheese came from New Zealand. The new industry meant not only economic prosperity for New Zealand that lasted until the 1970s, but massive social change around the country. It created our first large-scale industrial plants – the freezing works – thereby necessitating the invention of white gumboots. It also allowed for a proliferation of thousands of small-scale farms, including dairy farms, that could now be profitable. It is no exaggeration to say that Davidson and Brydone invented the New Zealand rural economy and created the very texture of our country and society.

Brydon and Davidson weren't the first to ship frozen meat, but they were there at the start, and are credited by many with founding the worldwide refrigerated meat industry. While the story that they invented refrigerated shipping is a myth, Brydon and Davidson are certainly New Zealand legends.

Mountain Buggy

I'M A MOUNT'N MAN AND
I LIKE MOUNT'N BUGGIES

1

At first they were bought mainly by people in rural areas and other runners, but they soon caught on with the urbanites who began to see the Mountain Buggy – much like their four-wheel-drive car – as a statement of what they aspired to be...

Right: One of the later models of the Mountain Buggy with a moveable front wheel.

In 1992, inventor and father Allan Croad was thinking about how to take the kids off his wife's hands and push them around while he was out running.

The baby carriage was invented in 1733 by English architect William Kent for the third Duke of Devonshire's children. An American inventor, William Richardson, patented a series of improvements to the baby carriage in 1889 which are seen as the beginnings of the baby buggy. Mountains were invented sometime around 4 billion BC, when the movement of the tectonic plates crashing against each other caused an overlap of land to stick up in the air. The name of the inventor is disputed.

Meanwhile, in 1992, inventor and father Allan Croad was thinking about how to take the kids off his wife's hands and push them around while he was out running. As a PE teacher he enjoyed exercise but with young kids he was somewhat hampered. Browsing an American magazine, he came across a design for a three-wheeled kids' buggy which he thought was okay, but could be improved on. In his garage he pulled together a prototype from an old kids' carseat and a golf trundler.

The initial tests were positive so he got in his car and drove up and down the Hutt Valley where he lived, trying to convince sceptical engineering firms to work with him to build a full system. At the time, he recalls, it was difficult to get people to realise he was anything but a crackpot inventor – but then he struck a bit of luck. A Wellington newspaper put a photo of him running with his invention on the front page, and the phone started ringing. The name came fast – he wanted to give the idea that this was something that could go everywhere a mountain bike could. He adapted the designs from overseas to add his own innovations that would better suit New Zealand conditions – a lightweight aluminium frame that could fold up, a hammock system to support the child, and knobbly tires that were smaller than the US version.

True to the name, at first they were bought mainly by people in rural areas and by other runners, but they soon caught on with the urbanites who began to see the Mountain Buggy – much like their four-wheel-drive car – as a statement of what they aspired to be, rather than who they currently were. The product took off, the brand gained awareness quickly, and the Mountain Buggy company was on a serious growth curve. The buggies were marketed and sold internationally and at first things went well.

Some tricky growth issues meant the company was stretched for growth capital, and Croad sold the business in 2004 to a larger company, Tritec. They proceeded to go into liquidation but were bought by their New Zealand competitor, Phil & Ted's Most Excellent Buggy Company Ltd. Phil & Ted's, to their credit, recognised the value of the brand and have retained the Mountain Buggy name, which is still respected worldwide.

Croad went on from Mountain Buggy to work on other innovations, including the LeisurePod – a camping system which fits on the back of a standard trailer and allows people to venture into the great outdoors in a quick and easy way.

While Croad may not have invented mountains or buggies, or even the idea of a buggy that could go into the outdoors, his innovative spirit improved the initial idea, and his product became world renowned as a real Kiwi innovation.

Trench Warfare

AKE, AKE, KIA KAHA E!

Final Campaign Against Titokowaru

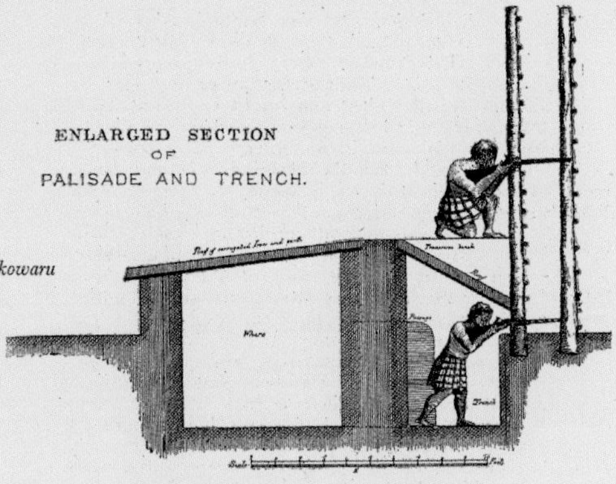

Section of Entrenchments, Tauranga-ika Pa

60

Hone Heke and Kawiti won (or at least tied) the Northland War in the mid-1840s by building, successfully defending, then annoyingly abandoning impressive pā at Ohaeawai, Puketutu and Ruapekapeka. Under musket, cannon and even rocket fire they suffered only 60 losses to the imperial forces' 300.

Right: Tauranga-Ika was perhaps the most formidable pā constructed in New Zealand, used to protect territory and as a base from which to launch raids against the Whanganui hinterland.

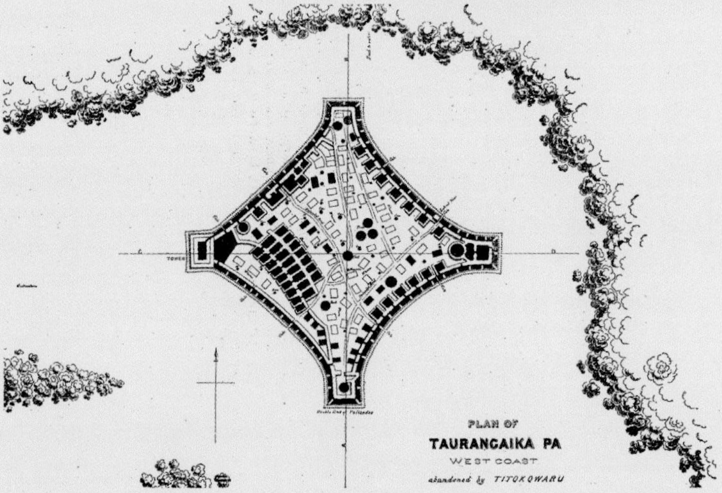

From a survey by J. Buchanan, 1869]

Plan of Tauranga-ika Pa, West Coast

As Kiwis we take great pride in inventing something that has a profound effect on the world. It doesn't seem to make any difference if it's a good thing or a nasty thing. If we'd invented the plague or cockroaches, we'd be in a pub in London yelling with ghoulish pride, 'Listen mate, WE invented that!'

Which is exactly what we say about trench warfare.

Pre-European Māori certainly fought using earthworks and fortifications. Their pā were elaborately defended with palisades, trenches, fences and pits. Rather than being the usual residence of an iwi, they were defensive structures that could be retreated to in the case of a threat. As well as residential whare they housed wells, food storage pits, and plantations.

Defensive earthworks on a pā could be extremely sophisticated and often included many concentric palisades (fences of stakes) lining the tops of ramparts behind which were networks of ditches, terraces and underground tunnels for communication and escape. Māori usually built pā on hills, headlands, ridges and islands to add elevation to the advantages of the earthworks.

After contact with the Europeans, pā proved well adapted to musket fighting, and the better armed colonial forces often came off second best when taking on pā. This defensive advantage was increased further when expert warriors like Ngāpuhi's Te Ruki Kawiti modified and improved traditional techniques and built pā especially for the purpose of frustrating the colonials.

Hone Heke and Kawiti won (or at least tied) the Northland War in the mid-1840s by building, success-

Above: Soldiers of the Māori Pioneer Battalion take a break from trench improvement work near Gommecourt, France, July 1918.

fully defending, then annoyingly abandoning impressive pā at Ohaeawai, Puketutu and Ruapekapeka. Under musket, cannon and even rocket fire they suffered only 60 losses to the imperial forces' 300. They had successfully adapted traditional pā fortifications to withstand the attacks of large numbers of soldiers with cannons and muskets. At Ruapekapeka, Kawiti and his men took shelter in dark underground bunkers while the shells rained down – giving the pā its name, meaning 'bat's nest'. It was certainly a form of warfare that was new to the imperials and they took careful notice of these successful defences. Major Mould (his real name) of the Royal Engineers surveyed the pā and took a scale model back to England where some say it was tabled in the House of Commons.

At the same time as Māori were making this response to imperial firepower, warfare was changing worldwide. The deadly accuracy of weapons and the speed of reloading had advanced quicker than the speed troops could move – making soldiers out in the open sitting ducks. Armies had used earthworks since armies were invented, but these methods became an increasing necessity and can be seen in notable examples in the 1700s (such as the American Revolution) and early 1800s.

Trench warfare on a large scale really began with the Crimean War in the 1850s, the American Civil War in the 1860s and the Second Boer War of 1899–1902 and, of course, reached its awful zenith with World War I. It didn't become obsolete until tanks and bomber aircraft were brought to the battlefield, making digging in pointless.

This defensive advantage was increased further when expert warriors like Ngā Puhi's Te Ruki Kawiti modified and improved traditional techniques and built pā especially for the purpose of frustrating the colonials.

There is no doubt that the British experience fighting the likes of Te Kawiti and Hone Heke in their trenches pre-dated those major European trench conflicts. Whether Māori fortifications had much influence on modern trench warfare, or whether World War I-style trench warfare had other parents closer to home is debatable.

Whatever the case, despite the fact that trench warfare – the eternal symbol for the futility of combat – is a strange and poisonous orphan for us to adopt, the myth that it was invented here is repeated so often that it is commonly held to be true. If we want to wrongly take the credit for something, I vote we choose something nice instead, like ice cream or rainbows.

Jandals and Ugg Boots

FORGET FOOTWEAR, THIS IS FOOTWAR

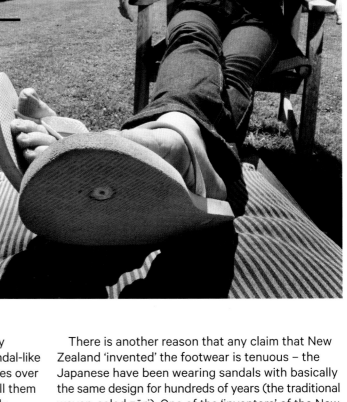

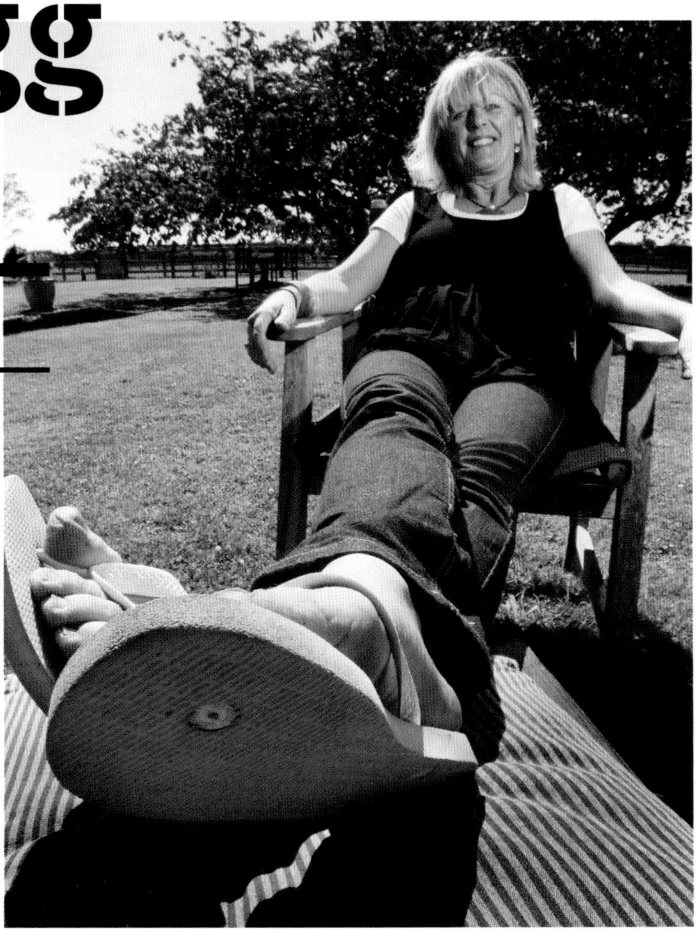

1957

For years it's been thought that one Morris Yock coined the phrase 'jandal' as an amalgam of 'Japanese' and 'sandal' after seeing similar things in Asia, then took out a trademark for the name in 1957.

Right: Cowie's daughter, Mary Deken, seen here modelling a pair of 'jandals' at her Taranaki home.

Firstly, let's acknowledge that we are the only country in the world to call the rubberised sandal-like footwear so common on New Zealand beaches over summer, 'jandals'. Other less refined souls call them flip-flops, or in Australia, thongs – presumably because they look like G-string pants for your feet. And there is a good reason for this – the word jandal is actually a trademark, and for years was owned by Skellerup, a New Zealand company with a Danish name, before manufacturing was moved to Malaysia and the brand was sold to Sandford Industries. They've had to defend the trademark and the New Zealand High Court is considering whether 'jandal' can actually be a trademark any more. As you can see, things are already looking shaky here.

There is another reason that any claim that New Zealand 'invented' the footwear is tenuous – the Japanese have been wearing sandals with basically the same design for hundreds of years (the traditional woven-soled zōri). One of the 'inventors' of the New Zealand sandal even acknowledged that it was the sight of a US businessman wearing rubber versions of the Japanese sandal that made him want to manufacture something similar in New Zealand.

However, the full tale of the New Zealand jandal is worth telling as the footwear is so much part of our culture now, and there is a great controversy and inter-family rivalry about the genesis of the product. For years it's been thought that one Morris Yock of Auckland coined the phrase 'jandal' as an amalgam

Left: A 23-year study of Northland beaches found that 70 per cent of washed-up jandals were left-footed.

To add to the mystery, both Yock and Cowie have passed on, and it turns out they were friends, so the debate is clouded, confused and slightly rubbery smelling.

of 'Japanese' and 'sandal' after seeing similar things in Asia, then took out a trademark for the name in 1957. At the time, he wasn't allowed to import them from Asia, so his sons started manufacturing them in a Te Papapa garage, and people liked them.

Lately, though, the family of one John Cowie have claimed that it was he who came up with the idea for both the product and the name after a brainstorming session with a bunch of Japanese businessmen. He was even manufacturing them in Hong Kong for a while in the late 1940s. To add to the mystery, both Yock and Cowie have passed on, and it turns out they were friends, so the debate is clouded, confused and slightly rubbery smelling.

But it seems their families can't handle the jandal, and the disagreement between them erupted when Surf Lifesaving New Zealand started using jandals as a fundraiser in the mid-2000s. Luckily no one appears to have been hurt so far in the long-running jandal scandal – a dispute that seems a little pointless. Ultimately, both should be acknowledged for their roles in bringing the footwear to New Zealand (onya boys!), but we all have to admit that New Zealand cannot claim the jandal as one of ours – it really is the Japanese sandal. And let's not forget, the real enemy is out there, with the Brazilian-made Havaianas brand now the dominant force in the global Japanese zōri-inspired footwear market.

Ugg Boots are another piece of popular footwear that some try to claim for New Zealand. Notwithstanding the other facts, why would we even *want* to claim a shoe that makes you lose IQ points just by picking them up? However, setting aside those admittedly subjective opinions, 'Ugg' is a trademark of an Australian company (imaginatively also called UGG). To get around this little legal point, Ugg Boots are also known as 'Ug Boots', 'Ugh Boots' and even 'Uggies', but what everyone agrees on is that the Ugg sound comes from the word 'ugly'.

Sheepskin boots have been made and worn in rural Australia for nearly 200 years, but the name only came in the late 1950s. They were worn by pilots during both world wars to keep their extremities warm in unpressurised cockpits ('Fug boots'), and the boots grew in popularity in the 1970s and '80s as a symbol of youth rebellion, very popular with the surfing community. The real sales kicked in when the UGG company started getting actors and actresses on popular US movies and TV shows to wear them – Pamela Anderson was seen sporting them for a while until she realised the fluffy stuff was wool and they were made from sheepskin; she then eschewed them as breaching animal rights. Forget animal rights, they are an assault on a human's right not to wear footwear that makes you look like a dag. Luckily for New Zealand, and bucking the usual trend, with 200 years of manufacturing to point to, this is one invention that Australia claims which they can have – Ugg Boots, and all their satanic variations, are Australian.

Kiwi Food

OUR JUNK FOOD INNOVATIONS

If you're anything like me, you rank junk foods such as cheese-flavoured corn snacks with the computer and the motor car as some of the most important inventions of the modern world. What hollow existences we would lead without Twisties, without the Moro or without Minties. How impoverished would our culture be if there were no Eskimos to unite us?

One thing that is quickly apparent when you travel overseas is that different countries celebrate different sets of junk foods. Some are universal, such as Coca-Cola, but most are regional. But are the junk foods of New Zealand really native foods, invented here, or are they imported and named to suit us? It may indeed be clutching at straws (or at least at potato sticks) but this is a round-up of which of our cherished brands are really ours and which are foreign.

Let's start with a couple of disappointments – and before I go on let me say that national pride must take a back seat to truth, at least in the following few paragraphs. Marmite and Vegemite are not New Zealand products at all. Marmite was invented in the UK and first imported to New Zealand by the Sanitarium company around the time of World War I. Vegemite was invented by an Australian pharmacist in 1922 (he was mixing brewers' and bakers' yeasts to try and make a better yeast for brewing). If you still want to eat it, that's your decision.

Hokey pokey ice cream is OURS! Other countries have the confection that we call hokey pokey. It is simply caramelised sugar, given little bubbles by the addition of baking soda. Honeycomb, humbug and butterbrickle are all hokey pokey-like, but nowhere else is it put into ice cream. In America, rocky road ice cream has lumps of stiff toffee in it, but it's not the same stuff at all. Hokey pokey ice cream was patented by one William Hatton from Dunedin in 1896, and made during World War II by Meadowgold Ice Cream. However, it's unlikely Hatton was the original creator, with evidence dating back to 1892 for a confection called hokey pokey. Since Hatton, though, the recipe has changed a little – originally, the hokey pokey was made in big sheets and smashed up with hammers, so that the hokey pokey lumps were rough with sharp edges. Now the hokey pokey is made in roundish pellets – but it's all good. We're exporting globs of the stuff to Japan and the Pacific, and all together in New Zealand, somebody is eating about 5 million litres of hokey pokey in a year. You know who you are, chunky.

We're exporting globs of the stuff to Japan and the Pacific, and all together in New Zealand, somebody is eating about 5 million litres of hokey pokey in a year. You know who you are, chunky.

Do you think of Weet-Bix as being essentially New Zealand? Well, once again it is Australian. Even the song, 'Kiwi kids are Weet-Bix kids' was originally 'Aussie kids are Weet-Bix kids'. When I first heard that jingle on television in Australia, I almost choked. Looking back I was naïve to be fooled; if you look closely, the kids in the ads are obviously Australian. Again, if you still want to eat Weet-Bix, go ahead. It's up to you.

With cheese-flavoured corn snacks, there is good news and bad news. The technology for making the snacks was developed overseas. The machines that make them were developed from the machines that spun the foam for inside car seats. (If you still want to eat them…) This technology was brought to New Zealand in about 1965 and we began to make our own sorts.

Rashuns are the most New Zealand cheese-flavoured corn snack. Nowhere else in the world are Rashuns made. The recipe (the shape and the flavour) was invented here, and they are exported to some Pacific islands. The name comes from a combination of 'rations' and 'rashers' for the bacon flavour (now *that's* Kiwi ingenuity!).

Having been made now for 35 years, Twisties were the first New Zealand corn snacks and the next most Kiwi. According to the manufacturer, the 'puffy worm shape' is probably not ours, and the flavour isn't either, and neither is the name (there are Twisties in Australia but they are a different snack), but the combination is all ours, and like Rashuns they are exported to French Polynesia. They are about five pack designs behind us, but as the brand manager says, 'That's the French for you.'

Other snacks are not at all Kiwi, but are made under licence from overseas companies – these includs Cheezels, Burger Rings, Big-Uns, Cheese Balls, Poppa Jacks and Munchos.

We've managed to come up with quite a few successful chocolate bars, but none that have taken the world by storm. The Moro is our best known bar by far. The recipe is New Zealand, probably based on the overseas Mars bar, and first made in 1968. The Crunchie doesn't belong to us, it's British. The Pinky was invented in New Zealand in the 1960s. Buzz Bars and the Perky Nana are also uniquely ours; the Buzz Bar goes back to the 1930s. Cadbury chocolate bars (Dairy Milk, Caramello, Energy, etc.) are available all over the world. Every country occasionally brings out a new flavour that lasts a while then disappears. There is one ray of light. New Zealand has spawned one flavour of Cadbury chocolate which has persisted, and now is an international favourite. In the early 1990s the Dunedin chocolate researchers came up with putting cherries and bits of biscuit into chocolate bars, and New Zealand gave the world Black Forest. (And then named it after a German forest. You idiots! What was wrong with Kaingaroa?)

But as any wind-blooded Wellingtonian will tell you, the *real* local chocolate comes from Whittaker's, whose Peanut Slab is the stuff of legend and much sought after by homesick expat Kiwis eking out their existence by pulling pints in some London pub – them, and Maggi onion soup dip, that ambrosia concocted from dry soup powder and reduced cream. It tastes so much better than it sounds.

New Zealand has a soft drink invention of our very own which has survived. Lemon & Paeroa is now owned by the Coca-Cola company, but is marketed only in New Zealand. It all began in 1904 when Dr Arthur Wollman tasted some mineral water bubbling from a spring in a cow paddock in the Hauraki Plains area. He thought it was good and he should have known – he was the government balneologist. In those days the government had balneologists – scientists whose field was the medical benefits of mineral waters – on their payroll. This was a very important matter for early New Zealand because our major tourist attraction in the 19th and early 20th centuries was our mineral waters. People came from all over the world to bathe in the warm springs and drink our spring water.

In 1907 a company called Menzies & Co. bought the access to the water and began to bottle it. They mixed it with lemon and called it Paeroa & Lemon. Their major market was in the city of Auckland, and when it began to get expensive to transport the water there, they copied Wollman's chemical analysis of the water exactly and thus made a synthesised version. They reversed the name to Lemon & Paeroa, and it has continued to sell to this day. Paeroa's neighbouring town, Te Aroha (which, if you have never been there, is beautiful!) also at one stage had a soft drink named after it. Lemon & Te Aroha (if that is what it was called) has not survived, and the people of Paeroa have been unbearable ever since, erecting a huge brown L&P bottle on their main street.

But the best stories in the world of confectionery may be lost to us forever. The process of enrobing shaped marshmallow in chocolate may be a New Zealand invention. If it isn't, many of the resulting candies certainly are. The records of Cadbury and Hudsons cannot confirm it, but it is thought that the chocolate fish is a uniquely New Zealand confection invention. The marshmallow chocolate Easter egg, too, may be uniquely our invention. Both appeared around 1955 as novelties at Cadbury's. Minties and Jaffas and Snifters belong to Australia but Fruit Bursts (those chewy things wrapped like Minties) were invented here and have gone on to conquer Australia.

The most interesting thing that comes out of all this is that companies who manufacture these foods have people researching and designing foods, snacks, bars and lollies. If the lollies sell, they stay, but either way they remain just recipes to the companies that make them. That's why we retained Goody Goody Gumdrops, thanks to the engineering excellence of the tip-top team at Tip Top, but have unfortunately lost Robin Hoods, and the only Joy Bars known to youth today are not as delicious as they used to be.

No. 8 Wire

OUR IMPORTED SYMBOL OF INVENTION

8

It wasn't until 1963 that New Zealand Wire Industries (now Cyclone) began to make wire in New Zealand. High-tensile wire soon replaced the old steel wire, and No. 8 gave way to the new 2.5mm, which is lighter, cheaper and stronger.

Below: One small part of the millions of metres of No. 8 Wire stretched across Godzone.

No. 8 Wire is legendary as the stuff of all New Zealand inventions. The term 'No. 8 Wire' is inextricably entangled with many of our icons of Kiwiness: Barry Crump, Fred Dagg, self-sufficiency and 'she'll be right'. It is ironic, weird and a little bit disappointing, then, that No. 8 Wire is itself an import.

No. 8 Wire is actually more properly called 'ISWG#8' – a clumsy acronym that stands for the British 'Industry Standard Wire Gauge 8'. Apparently different kinds of wire are measured and named by their thickness or diameter, with the thickest called zero gauge (at an impressive half an inch, or 10mm thick, No. 0 wire is not so much wire as a rod). It seems 7 (5mm) and 8 (4mm) gauge are the easiest to handle, and therefore are the most popular for fencing throughout the world.

It was in the 1860s that New Zealanders began importing wire for fencing. In those days it was made of pure iron, but by the 1880s steel had replaced the rusty iron as the wire makers' metal of choice. Eight-gauge wire was the most popular and became ubiquitous in New Zealand; our country was, if not covered by it, at least divided up into tiny pieces by it, and every single inch (or centimetre) of it was imported from England, Belgium, Germany or Australia.

It wasn't until 1963 that New Zealand Wire Industries (now Cyclone) began to make wire in New Zealand, putting an end to a century of wire importing. High-tensile wire soon replaced the old steel wire, and No. 8 gave way to the new 2.5mm, which is lighter, cheaper and stronger.

Because No. 8 Wire was available at all times around the farm, it was put to millions of uses – to tie, bind, construct, repair, modify and fashion things. It was often used to make the prototypes of real innovations, later manufactured properly. Number 8 wire came to stand for adaptability. In fact, at the annual Fieldays near Hamilton, they have a competition for making new creations utilising good old No. 8. There are even rumours that many of the All Blacks who played off the back of the scrum at number 8 were in fact fashioned out of wire.

As flexible as No. 8 Wire was, as many things as it has been borrowed to make and as many New Zealand inventions for which it has helped form the initial prototype, No. 8 was eventually found wanting in one important area: fence making. It took Kiwi William Gallagher to invent a better fence (see page 234), perhaps the final irony for the product – and in doing so, Gallagher provided us with proof that No. 8 can become Re-wired.

Thanks

Brian Sweeny of NZEdge.com, one of New Zealand's best online assets, for his generosity and sharing his thoughts, research and time
Shaun Hendy, professor at the University of Auckland and co-author of *Get off the Grass*
Joshua Downs, who did research, editing, and some writing for us
Honey E, a freelance researcher from the Philippines whom we found on guru.com
Will Seal, not a guarantee, but a researcher mate of Jon's
Neville Queree and Justin Brownlee and the scientists from Callaghan Innovation (aka IRL, DSIR)
Simon Martin, of Hudson Gavin Martin, Technology and Intellectual Property Lawyers
Dave Winsborough, Director, The Winsborough Group for his insights on leadership
Greg Shannahan of the Technology Investment Network (TIN)
The staff of New Zealand Trade & Enterprise who shared ideas
Dr Sean Simpson, CEO of Lanzatech
Dr Ian Taylor, CEO of Animation Research
Treasure Loader, for helping create the book's companion website – www.no8rewired.co.nz
Sir Peter Gluckman, Chief Science Advisor to the Prime Minister
Professor Hamish G. Spencer, Director of the Allan Wilson Centre
Alan Sharp, for use of his unpublished dairying heritage research
Jack Downs, for image research
Scott and Cerian Hamilton, for thoughtful inspiration
Jim Stewart, Shirley Walker and Marilyn Murdoch, for posting precious photographs to us
Dr Brian A. Tinsley, Professor Emeritus, University of Texas at Dallas
New Zealand Story, http://story.newzealand.com
All the inventors, innovators and enthusiasts who shared their stories and their time

Bibliography

Get Off The Grass, Shaun Hendy and Paul Callaghan, Auckland University Press, 2013
101 Ingenious Kiwis, Tony Williams, Reed Publishing, 2006
The Littlest Clue, Selwyn Parker, Industrial Research Limited, 2002
Kiwi Ingenuity, Bob Riley, AIT Press, 1995
TIN100, The Technology Investment Network report, 2013
New Zealand by Design, Michael Smythe, Random House, 2011
The Penguin History of New Zealand, Michael King, Penguin Books, 2003
Te Ara: The Online Encyclopedia of New Zealand, www.teara.govt.nz
Cracking the Einstein Code: Relativity and the Birth of Black Hole Physics, Fulvio Melia, University of Chicago Press, 2009
NZ Edge, the online celebration of the global life of New Zealanders, www.nzedge.com
Papers Past, online collection of historical newspapers from the National Library of New Zealand, http://paperspast.natlib.govt.nz

Image Credits

Page 4 Chris Williams/NZ Story **6** Damian Alexander/NZ Story **10** David Wall/NZ Story **12** Martin Aircraft Company **15** Corbis (BE062043) **16** Rocket Lab Ltd **17** *New Zealand Herald* **18** Keith Alexander **19** Peter Lynn **20** G. R. 'Dick' Roberts, © Natural Sciences Image Library **21** Alexander Turnbull Library (1/2-106439-F) **22** Director: Nathan Price, Studio: Ruskin Film, Agency: Colenso BBDO, Client: State Insurance **23** Alexander Turnbull Library (WA-42009-F) **24–25** Chris Williams/NZ Story **26** NZ Story **28–29** TruTest **30** EasiYo Products Ltd **31** Light family **32** Revolution Fibres **33** Three Over Seven **34** Te Ara – The Encyclopaedia of New Zealand **36** Black Forest Park **37** Murdoch Family **38–39** Massey University Archives **40** Alexander Turnbull Library (Eph-A-DAIRY-1903-01-front) **42** Allflex **44** Baker No-Tillage Ltd **45** Kelly Tarlton's Sea Life Aquarium **46** Archives NZ (AAQT 6539/A1919) **48–49** Murdoch Family **50** Chris Williams/NZ Story **52–53** Rex Bionics **54** Jake Evill **55** Murdoch Family **56** Alexander Turnbull Library (1/2-C-10320-F) **57** Alexander Turnbull Library (PAColl-6301-18) **58** Auckland Bioengineering Institute, University of Auckland **59** StretchSense Ltd **60** Wikipedia/Creative Commons **61** *Washington Post* **62–63** Pacific Edge **64** (L) Garth Sutherland/Smartinhaler, (R) Mesynthes Ltd **65** D. Durand/Children's Hospital and Research Center, Oakland, CA **66** *Auckland Star* **67** New Zealand Post **68–69** Institute of Environmental Science and Research **70–71** Fisher & Paykel **72** Damian Alexander/NZ Story **74** Alexander Turnbull Library (Eph-A-DANCE-1926-02) **75** Wikipedia/Creative Commons **78–79** South Canterbury Museum (4963) **80** Alexander Turnbull Library (1/1-020462-G) **82** Alexander Turnbull Library (B-185-003) **83** Fonterra **84** Alexander Turnbull Library (EP/1953/0034-G) **85** Archives NZ (AAQT6538/1) 86 Wikipedia/Creative Commons **87** Papers Past/National Library of New Zealand **88** Tourism New Zealand/NZ Story **90** HamiltonJet **92** Laurie Callender **93** Jon Bridges **94** Yike Bike **96–97** Sealegs Corporation Ltd **99** Armour Transport Technologies Ltd **100** Te Ara – The Encyclopaedia of New Zealand **101** Shirley Walker **102** Romotow Holdings Ltd **104** Te Ara – The Encyclopaedia of New Zealand **105** Wikipedia/Creative Commons **106** Harry Ruffell, www.brittendvd.co.nz **109** Terry Roycroft **110–11** Gibbs Sports Amphibians Inc. **112** Destination Rotorua/NZ Story **114** www.sciencephoto.com (H418/0094) **116** Alexander Turnbull Library (EP/1974/3501-F) **118–19** *Scientific American* **120** Wikipedia/Creative Commons **121** Zespri **122** Eric Adank **123** Archives NZ (AAME 8106 W5603 101 9/11/20) **124** Dental School, University of Otago (MS-2616/063) **125** Dental School, University of Otago (AG-508/83) **126–27** Katherine Downs **129** David Downs **130** Library of the London School of Economics and Political Science (LSE/IMAGELIBRARY/6) **132** Chris Williams/NZ Story **134** Canterbury Museum (S6/1; photograph by H. H. Clifford) **136** Wikipedia/Creative Commons **137** Allan Wilson Centre, Massey University **138** David Downs **139** Amanda Buckingham **140** Toby Downs **141** Rockflower **142** Wikipedia/Creative Commons **143** Professor Brian A. Tinsley **144** Richard McCulloch **145** QP Sport **146** Alexander Turnbull Library (114/275/20-F) **148** Alexander Turnbull Library (PAColl-6303-51) **149** *New Zealand Herald* **150** Hutch Design Ltd, www.kindlingcracker.com **151** D'Arcy Polychrome Ltd **152** Chris Williams/NZ Story **154** WilliamsWarn **156** Wikipedia/Creative Commons **157** Photosport (PS0017728) **158** Wikipedia/Creative Commons **160** Crowther Audio **161** Fastmount **162–63** Animation Research Ltd **164** Wikipedia/Creative Commons **165** Geoff Barnett **166** *New Zealand Herald* **168** Chris Williams/NZ Story **170–71** National Aeronautics and Space Administration (NASA) **172** Electronic Navigation Ltd **173** Callaghan Innovation **174–75** Vaughan Jones **176** www.sciencephoto.com (C003/5501) **177** Wikipedia/Creative Commons **178–79** Lanzatech **180** Kode **182–83** Mark Poletti **186** Wayne Simpson, AgResearch **187** AgResearch **188–89** Alistair Campbell **190** Richard Faull **191** Augusto Ltd **192–93** Comvita **194** Wikipedia/Creative Commons **196** Chris Williams/NZ Story **198** Alexander Turnbull Library (PA1-q-228-01) **199** *New Zealand Herald* **201** David Downs **202** Patent Application – Edlin/Stewart **204** Smitkin Marketing Ltd **205** Tim Cox **206** Owaka Museum **207** Wikipedia/Creative Commons **208** Peter Davies **210** www.substech.com **212** Wikipedia/Creative Commons **214** Qualcomm Halo **216–17** PowerbyProxi **218–19** GCSuperconductor **220** New Zealand National Fieldays Society **221** Hamish Scott **222–23** ArcActive **226** Fisher & Paykel Appliances **227** Vinny Lohan **228** Magritek **229** Department of the Prime Minister and Cabinet (DPMC) **230** Robinson Seismic **232** Chris Williams/NZ Story **234** Gallagher Group **236** Grant Pearce **237** Fenceworld Ltd **238** Patent Application – John Reid **239** Alexander Turnbull Library (1/2-022050-G) **240** Roland Mathews **241** Vega Marine Aids **242–43** Claudio Petronelli **244** BFM Fitting **245** ECHO Reusable Envelope **246–47** Cavotec MoorMaster Ltd **248** Securichain **249** Kaynemail **250** Goodnature Ltd **251** Precision Seafood Harvesting **252** Chris Williams/NZ Story **254** Te Ara – The Encyclopaedia of New Zealand **255** Zespri **256** Auckland Art Gallery Toi o Tāmaki (1978/34, purchased 1977) **258** University of Otago, Hocken Collection (02-1) **259** Alexander Turnbull Library (1/1-009113-G) **260–61** Alan Croad/Mountain Buggy **262** James Cowan, *The New Zealand Wars: A History of the Maori Campaigns and the Pioneering Period* (vol. 2), Wellington: Government Printer, 1922, **p.289** **263** Alexander Turnbull Library (1/2-013414-G) **264** Mark Dwyer/*Taranaki Daily News* **265** Melanie Lovell-Smith **268** Pauline Woodcock

Index

Illustrations are indicated by *italic* type

Jon Bridges

Jon is a well-known writer, TV presenter and producer, appearing in *IceTV*, *Who Ate All the Pies?*, and the panel show *Would I Lie to You?*. For the past five years he has been the producer of New Zealand's most successful TV comedy show, *7 Days*, which he co-created.

A sought-after public speaker and writer, Jon has written a regular column in the *New Zealand Listener*, a popular blog *We're Building a House* on www.stuff.co.nz, and a book on cycling in New Zealand, *Easy Rider*. Jon also co-wrote three other books with David, including the precursor to this one, *No. 8 Wire*, in 1998.

Jon lives in Three Kings, Auckland, with his furniture designer wife Gemma and their recent addition Zeno.

David Downs

A native of Whanganui, David began his working life as an actor, writer, comedian and co-founder (with Jon and others) of The Classic Comedy & Bar in Auckland. He left this behind to start a second career in IT, which he stuck with for nearly 20 years, including a stint living and working for Microsoft in Asia. David's strong interest in innovative companies, and his pride in New Zealand, lead him back to Aotearoa and his third career at government agency New Zealand Trade & Enterprise, where he helps local businesses grow internationally.

David lives in Devonport, Auckland, with his TV producer wife Katherine and his three fast-growing boys Jack, Joshua and Toby.

PENGUIN BOOKS

Published by the Penguin Group
Penguin Group (NZ), 67 Apollo Drive, Rosedale, Auckland 0632, New Zealand (a division of Penguin New Zealand Pty Ltd)

Penguin Group (USA) Inc., 375 Hudson Street, New York, New York 10014, USA
Penguin Group (Canada), 90 Eglinton Avenue East, Suite 700, Toronto, Ontario, M4P 2Y3, Canada (a division of Penguin Canada Books Inc.)
Penguin Books Ltd, 80 Strand, London, WC2R 0RL, England
Penguin Ireland, 25 St Stephen's Green, Dublin 2, Ireland (a division of Penguin Books Ltd)
Penguin Group (Australia), 707 Collins Street, Melbourne, Victoria 3008, Australia (a division of Penguin Australia Pty Ltd)
Penguin Books India Pvt Ltd, 11, Community Centre, Panchsheel Park, New Delhi – 110 017, India
Penguin Books (South Africa) (Pty) Ltd, Block D, Rosebank Office Park, 181 Jan Smuts Avenue, Parktown North, Gauteng 2193, South Africa
Penguin (Beijing) Ltd, 7F, Tower B, Jiaming Center, 27 East Third Ring Road North, Chaoyang District, Beijing 100020, China

Penguin Books Ltd, Registered Offices: 80 Strand, London, WC2R 0RL, England

First published by Penguin Group (NZ), 2014
10 9 8 7 6 5 4 3 2 1

Portions of *No. 8 Re-wired* were previously published in *No. 8 Wire* (Jon Bridges and David Downs, Hodder Moa Beckett, 2000).

Cover and internal design by areadesign.co.nz
Typeset in Calibre and Karbon Slab Stencil, fonts by klim.co.nz
Prepress by Image Centre Ltd
Printed in China by Leo Paper Products Ltd

ISBN 978-0-143-57195-7

A catalogue record for this book is available from the National Library of New Zealand.

www.penguin.co.nz